Biology JEOPARDY

Pamela Keef

WALCH PUBLISHING

The classroom teacher may reproduce materials in this book for classroom use only.
The reproduction of any part for an entire school or school system is strictly prohibited.
No part of this publication may be transmitted, stored, or recorded in any form
without written permission from the publisher.

1 2 3 4 5 6 7 8 9 10
ISBN 0-8251-5036-1
Copyright © 2004
Walch Publishing
P.O. Box 658 • Portland, Maine 04104-0658
walch.com

Printed in the United States of America

Contents

To the Teacher .... *vii*

Unit 1—Introduction to Biology 1

- Section 1: The Process of Science, Characteristics of Life I, Characteristics of Life II, Common Metric Units, Metric Unit Conversions 2
- Section 2: Microscopes, Branches of Biology I, Branches of Biology II, Branches of Biology III, The Composition of Matter I 4
- Section 3: The Composition of Matter II, Properties of Water I, Properties of Water II, Elements in Organic Compounds, Organic Compounds I 6
- Section 4: Organic Compounds II, Organic Compounds III, Organic Compounds IV, Organic Compounds V, Organic Compounds VI 8

Unit 2—Ecology 11

- Section 5: Levels of Organization, Obtaining Energy in an Ecosystem, The Flow of Energy in an Ecosystem, Ecosystem Relationships, Chemical Cycles 12
- Section 6: Terrestrial Biomes, Aquatic Ecosystems, Population Change, Ecosystem Change I, Ecosystem Change II 14

Unit 3—Cell Biology 17

- Section 7: Cell Discoveries, Cell Structures I, Cell Structures II, Cell Structures III, Cellular Organization 18
- Section 8: The Cell Membrane I, The Cell Membrane II, Photosynthesis and Energy Carriers, Photosynthesis I, Photosynthesis II 20
- Section 9: Photosynthesis III, Cellular Respiration I, Cellular Respiration II, Cellular Respiration III, Cell Reproduction I 22
- Section 10: Cell Reproduction II, Cell Reproduction III, Cell Reproduction IV, Cell Reproduction V, Cell Reproduction VI 24

Unit 4—Genetics 27

- Section 11: Mendelian Genetics I, Mendelian Genetics II, Mendelian Genetics III, Genetic Crosses I, Genetic Crosses II 28
- Section 12: Nucleic Acids I, Nucleic Acids II, Nucleic Acids III, Nucleic Acids IV, Nucleic Acids V 30

Section 13: Gene Expression, Genetic Errors, Inheritance Patterns, Human Inheritance I, Human Inheritance II 32

Section 14: Human Inheritance III, Human Inheritance IV, DNA Research I, DNA Research II, DNA Research III 34

Unit 5—Evolution 37

Section 15: Earth's History I, Earth's History II, Origin of Life, Evolution I, Evolution II 38

Section 16: Evolution III, Evolution IV, Evolution V, Evolution VI, Primates 40

Section 17: Human Evolution, Classification I, Classification II, Classification III, Classification IV 42

Unit 6—Microorganisms and Fungi 45

Section 18: Bacteria Classification, Characteristics of Bacteria I, Characteristics of Bacteria II, Characteristics of Bacteria III, Viruses I 46

Section 19: Viruses II, Protist Phyla I, Protist Phyla II, Protist Characteristics I, Protist Characteristics II 48

Section 20: Protist Characteristics III, Fungi I, Fungi II, Fungi III, Potpourri 50

Unit 7—Plants 53

Section 21: Plant Phyla I, Plant Phyla II, Mosses and Their Relatives, Ferns and Their Relatives, Plant Cells and Tissues I 54

Section 22: Plant Cells and Tissues II, Plant Cells and Tissues III, Roots, Stems I, Stems II 56

Section 23: Leaves, Flowers, Reproduction in Seed Plants I, Reproduction in Seed Plants II, Seeds and Fruit I 58

Section 24: Seeds and Fruit II, Plant Hormones, Plant Responses I, Plant Responses II, Potpourri 60

Unit 8—Invertebrates 63

Section 25: Characteristics of the Animals I, Characteristics of the Animals II, Characteristics of the Animals III, Characteristics of the Animals IV, Characteristics of the Animals V 64

Section 26: Invertebrate Classification I, Invertebrate Classification II, Sponges, Cnidarians and Ctenophores I, Cnidarians and Ctenophores II 66

Section 27: Flatworms I, Flatworms II, Rotifers and Roundworms, Annelids I, Annelids II 68

Section 28: Mollusks I, Mollusks II, Arthropod Characteristics I, Arthropod Characteristics II, Crustaceans I 70

Section 29: Crustaceans II, Chelicerates and Uniramians, Insects I, Insects II, Echinoderms 72

Unit 9—Vertebrates 75

- **Section 30:** Chordates and Nonvertebrate Chordates,
  Vertebrate Classification, Fishes I, Fishes II, Fishes III 76
- **Section 31:** Amphibians I, Amphibians II, Reptiles I, Reptiles II, Birds I 78
- **Section 32:** Birds II, Bird Orders, Mammals I, Mammals II, Mammal Orders 80

Unit 10—Human Biology 83

- **Section 33:** Tissues and Organ Systems I, Tissues and Organ Systems II,
  The Skeleton I, The Skeleton II, Bone Structure 84
- **Section 34:** Bone Formation, Joints, The Muscular System,
  Muscle Movement, The Integumentary System 86
- **Section 35:** The Digestive System I, The Digestive System II, The Digestive
  System III, The Excretory System I, The Excretory System II 88
- **Section 36:** Food and Nutrition, The Human Heart I,
  The Human Heart II, Blood Vessels, Blood 90
- **Section 37:** The Circulatory System and Blood Disorders, Blood Types,
  The Lymphatic System, The Respiratory System I,
  The Respiratory System II 92
- **Section 38:** The Nervous System I, The Nervous System II,
  The Neuron, The Brain, The Senses I 94
- **Section 39:** The Senses II, Drugs and the Nervous System,
  The Endocrine System, Endocrine Glands, Hormones I 96
- **Section 40:** Hormones II, Endocrine System Disorders, The Male Reproductive
  System, The Female Reproductive System, The Menstrual Cycle 98
- **Section 41:** Fertilization and Development I, Fertilization and Development II,
  The Immune System I, The Immune System II, Cancer 100

Answer Key 103

To the Teacher

Biology Jeopardy deals with facts and principles related to the study of life. It is designed to be used for several purposes: as a fun and easy way to reinforce what is being studied, as a study guide, and as a review of the unit or a culminating activity. It challenges your students to remember important facts and encourages them to enjoy themselves in the process.

The format of *Biology Jeopardy* is like the popular television quiz show *Jeopardy!* A student is given the answer and is asked to provide the question. The fact given as a question is actually stated ("An organism made of only one cell"), not asked. The student response is given as a question ("What is a unicellular organism?"). Many students will be familiar with the format from watching the *Jeopardy!* television show or playing its computer or board-game versions.

The questions are classified according to general topic and further by category. This format lends itself to use with a variety of attention-keeping games. Some games are suggested here; you and your students may invent others.

A number of questions throughout the book have more than one correct response. Often, an alternate response is written in parentheses, for example, "What is a pollen cone (or male cone)?"

How to Use This Book

Each section, or game, consists of five general categories. Within each category are five questions, each assigned a point value of 5 through 25 depending on its relative difficulty, plus a bonus question. The bonus question is not necessarily more difficult; it may refer to an unusual fact or a less important one. It may be used in whatever way seems suitable. A point value of 5 for each bonus question would give the entire game 400 points; a value of 25 would make it a 500-point game.

These questions in this format may be used to play a variety of games. However, it may prove effective to allow the students an opportunity to find the answers to, or study, the questions first. You may wish to reproduce the questions for a series of assignments, and then use a game as an evaluation, a further review, or a culmination of the unit. You may find that using the questions without a game is adequate. For these reasons, the answers are presented separately at the back of the book rather than with the questions.

Feel free to modify *Biology Jeopardy*. If you have stressed something in your class that is not included in this game, it is easy to add questions. Your students will quickly learn how to make questions for you in order to extend the game. You can also modify the questions to make them easier or harder to fit the needs of each particular class. Your class can play the same game more than once, which will help them remember material much more easily.

The same basic procedure can be used for playing any number of different games. Here are the directions for a typical *Jeopardy!*-style game:

- Write on the board the categories for the game to be played along with point values for each question.

- Divide the class into teams. Play begins when one student asks for a question from a given category with a given point value. For instance, the student might say, "I want the 10-point question from the 'Terrestrial Biomes' category."

- The game leader then reads the 10-point question from the requested category.

- Any student on the team may answer. The first person on the team to raise his or her hand is called on. (It may be the student who asked for the category to begin with.)

- If the answer is correct, record points for the team. The student who answered chooses the category and point value for the next question.

- If the answer is wrong, subtract the point value of the question from the team score. A student from the other team now has the chance to answer the question. Whoever answers the question correctly chooses the category and point value for the next question.
- If no one can answer the question, give the correct answer to the group. The student who last successfully answered a question chooses the next category and point value.
- When all the questions in the category have been used, erase the category from the board. Continue until all the categories are erased and the game is over.

Following are some other variations of the game:

Rounds

The categories and point values are displayed and the value of the bonus question is agreed upon. Bonus questions are not used until last. A scoreboard is drawn on the board to show the teams and what score they receive in each round.

The class is divided into three, four, or five groups, each having an equal number of students. (Up to 30 can play. Extra pupils may serve as scorekeepers, readers, or board keepers.) The players in each group or team sit or stand in a set order—first player, second, and so forth.

The game begins with Player 1 on Team 1 requesting a question. If the player responds correctly, the earned score is recorded under Team 1/Round 1. If the response is incorrect, the correct answer is read and a score of 0 is recorded. In either case, the point value is erased under the respective category. Then Player 1 of Team 2 has a turn to choose a question. After all the first players on each team have played, the play goes to the second players of each team, then the third, and so forth.

The game continues for as many complete rounds as possible. There may be several unused questions. If there are 30 players, the last player in each team chooses a category for a bonus question. Otherwise, the bonus question for each team is given to, or chosen by, the team's top scorer or chosen captain, either for that player or for the team to answer. The top-scoring team wins.

Progression

This game is set up like Rounds, preferably in five groups. The first players on each team choose a category for 5 points, the second players choose a question for 10 points, the third players go for 15, and so forth. Play continues for as many complete rounds as possible, with bonus questions handled as in Rounds.

Concentration

First, the categories and point values are written on the board and the bonus value is determined. The class is divided into two teams. The first player on one team requests a question. If the player replies correctly, his or her team gets the points, and the point value is erased below the respective category. If the player does not answer correctly, the response is announced to be wrong and nothing is erased from the board. The first person on the opposite team then chooses a question. The play goes from team to team, with each person choosing a question still listed on the board. The advantage goes to the person who knows the answer to a previously asked question and can remember where it is located on the board. Play continues until all questions have been used. The highest-scoring team wins.

Last Chance

The class is divided into two, three, four, or five teams, with the players seated or standing in a set order. The categories and point values are displayed, and the bonus value (perhaps generous) is chosen. The bonus questions are not used in regular play.

Player 1 on the first team requests a question. If the player replies correctly, his or her team earns the respective points; if the reply is incorrect, the teacher tells or explains the answer. In either case, the point value under that category is erased. The play then goes to Player 1 on the second team, who requests a question. After all the first players have had a turn, the play goes to the second players on each team, then the third, and so forth.

When all the questions have been used, the scores for each team are calculated. The next player on the lowest-scoring team chooses a category for the bonus question for his or her team. The teacher reads the question and accepts only one answer from the team. (The players may confer in order to come to an agreement.) If the reply is correct, the bonus score is added

to their total. Then the second-lowest-scoring team chooses a category, then the third, and the fourth, if there are that many teams. Only one bonus question is given to each team. There may be some that are not used. The winning team is that which has the highest score.

Solo

This game is played like Last Chance, except that it is played by five players instead of teams. The play goes from one player to the next in succession until all questions are used. Then each has a chance to choose a bonus question to raise his or her score. The top scorer wins.

Biology Bee

This game is played like a spelling bee, but no one is eliminated. First the categories and point values are displayed, and the value of the bonus question is determined. The class is divided into two teams. The first person on one team asks for a question by stating a category and point value. If the player responds correctly, his or her team receives the points and that point value is erased under that category. The next turn is taken by the first player on the other team, who chooses a question. However, if the first player's response is not correct, the same question is repeated for the first player on the other team. If the player replies correctly, his or her team gets the points and the play then goes to the second player of the first team. The play continues from one side to the other, with points going to the teams that answer correctly and the respective category points being erased from the board. The game is over when all 30 questions have been used. The team accumulating the most points wins.

No matter how you use *Biology Jeopardy*, it is an entertaining and stimulating way to review, and it's an excellent change-of-pace activity. You'll find your students eager to play it again and again

UNIT 1

Introduction to Biology

Introduction to Biology
Section 1

1

THE PROCESS OF SCIENCE	CHARACTERISTICS OF LIFE I	CHARACTERISTICS OF LIFE II	COMMON METRIC UNITS	METRIC UNIT CONVERSIONS
Uses one of the five senses to gather information about an object or event	Surrounded by a membrane; it is the basic unit of all living organisms.	Any environmental condition that causes an organism to adjust to a new condition	The SI base unit for mass	The number of milliliters in a liter
A collection of information gathered through observations	An organism made of only one cell	A reaction to an environmental condition	The SI base unit for length	The number of grams in a kilogram
A conclusion based on experience rather than direct observations	An organism made of many cells	A change over time among a group of organisms	The metric unit for volume	The number of centimeters in a meter

5

10

15

Biology Jeopardy

©2004 Walch Publishing

The number of deciliters in a liter	The number of millimeters in a kilometer	The number of centigrams in a decagram	
The SI base unit for time	The SI base unit for temperature	The SI base unit for electric current	
The sum of all the chemical reactions that take place within a cell or an organism to build up and break down needed materials	The process by which an organism maintains relatively stable internal conditions	A segment of DNA found on a chromosome that codes for a protein	
The combination of genetic material from two parents to create a new individual	The process in which an offspring receives all of its genetic material from one parent	The process by which an organism grows and changes over time	
A testable explanation for a set of observations	A hypothesis that has been well tested and repeatedly verified by experiments	A test that changes only one variable and keeps all other variables the same	
20	**25**	**BONUS**	**NOTES**

©2004 Walch Publishing

Biology Jeopardy 3

Introduction to Biology
Section 2

	MICROSCOPES	BRANCHES OF BIOLOGY I	BRANCHES OF BIOLOGY II	BRANCHES OF BIOLOGY III	THE COMPOSITION OF MATTER I
					A subatomic particle with very little mass and a negative electrical charge found outside the nucleus of an atom
1		The study of animals	The study of plants	The study of viruses	
					A subatomic particle with a mass of one atomic mass unit and a positive electrical charge found in the nucleus of an atom
5	An instrument that allows light to pass through a specimen and magnifies the image using two or more lenses	The study of mammals	The study of the interactions between organisms and their environment	The study of fungi	
10	On a compound light microscope, this part is where a specimen slide rests, and it allows light to pass through a specimen.	The study of ancient life and fossils	The study of microorganisms	The study of reptiles and amphibians	A subatomic particle with a mass of one atomic mass unit and no electrical charge found in the nucleus of an atom
15	Located on the nosepiece of a compound light microscope, these magnify the image of a specimen.				

4 Biology Jeopardy   ©2004 Walch Publishing

The number of protons in an atom	The study of butterflies and moths	The study of animal behavior	The study of birds	A powerful microscope that uses a beam of electrons to magnify a specimen
The sum of protons and neutrons in an atom	The study of the formation, early growth, and development of living organisms	The study of heredity and genetic variation among organisms	The study of fish	A microscope that passes a beam of electrons through a specimen to create an image
Atoms of the same element with different numbers of neutrons	The study of the structure of tissues	The study of insects	The study of cells	A microscope that passes a beam of electrons back and forth over the surface of a specimen to create a three-dimensional image

20

25

BONUS

NOTES

Introduction to Biology
Section 3

	THE COMPOSITION OF MATTER II	PROPERTIES OF WATER I	PROPERTIES OF WATER II	ELEMENTS IN ORGANIC COMPOUNDS	ORGANIC COMPOUNDS I
5	A pure substance made of only one type of atom	A combination of two or more substances that can be mixed together physically but do not combine chemically	An attraction between molecules of the same substance	An element with the symbol C	Single organic molecules that can be linked together to make larger molecules
10	A substance made of different elements that are chemically combined in a fixed ratio	A mixture in which one or more substances is dissolved and evenly distributed in another substance	An attraction between molecules of different substances	An element with the symbol H	A large molecule made by linking together many monomers
15	A group of two or more atoms that are held together by covalent bonds	The substance dissolved in a solution	The ability of water to move up narrow tubes against the force of gravity	An element with the symbol O	The chemical process of adding a monomer to another monomer or to a polymer by removing a water molecule

20	An atom that has gained or lost electrons and has an electrical charge	The dissolving substance in a solution	Any substance that forms hydrogen ions (H^+) in a water solution	An element with the symbol N	The chemical process of removing a monomer from a polymer by breaking a bond using a water molecule
25	A bond formed between two ions with opposite charges	The name given to a water molecule because charge is unevenly distributed between the oxygen and hydrogen atoms	Any substance that forms hydroxide ions (OH^-) in a water solution	An element with the symbol P	Molecules with the same chemical formula but different three-dimensional forms
BONUS	A bond formed when two or more atoms share electrons	The type of bond that holds two water molecules together	A solution made with a weak acid or base that can neutralize small amounts of strong acids or bases	An element with the symbol S	A type of bond found most frequently in carbon compounds
NOTES					

Introduction to Biology
Section 4

	Organic Compounds II	Organic Compounds III	Organic Compounds IV	Organic Compounds V	Organic Compounds VI
5	An organic compound containing carbon, hydrogen, and oxygen in a 1-to-2-to-1 ratio	A disaccharide made of two glucose molecules linked together	A type of polymer that contains more carbon and hydrogen than oxygen and does not dissolve in water	Large organic molecules that store genetic information	A term for a compound that speeds up the rate of a chemical reaction
10	A monomer of carbohydrate	A carbohydrate made of three or more monosaccharides	A fatty acid tail that contains all single bonds between carbons	A nucleic acid that contains the sugar deoxyribose	A type of protein that acts as a catalyst in organisms
15	The name of one of the two isomers of glucose	A polysaccharide made by plants to store energy	A fatty acid tail that contains only one set of double bonds between carbons	A nucleic acid that contains the sugar ribose	The amount of energy needed to start a chemical reaction

The starting materials in a chemical reaction	New compounds made in a chemical reaction	A reactant catalyzed by an enzyme	
A polymer made of amino acids	The number of amino acids that share the same basic structure	A type of covalent bond found between amino acids	
A fatty acid tail that contains more than one set of double bonds between carbons	A three-carbon alcohol that is the backbone of many lipids	The most abundant lipid in a cell membrane	
A polysaccharide made by animals to store energy in the muscles and liver	A polysaccharide found in plant cell walls	A polysaccharide containing nitrogen that is found in the exoskeleton of arthropods and in the cell walls of some fungi	
A carbohydrate made of two monosaccharides	A disaccharide made of glucose and fructose	A disaccharide made of glucose and galactose	
20	**25**	B O N U S	N O T E S

Biology Jeopardy

UNIT 2

Ecology

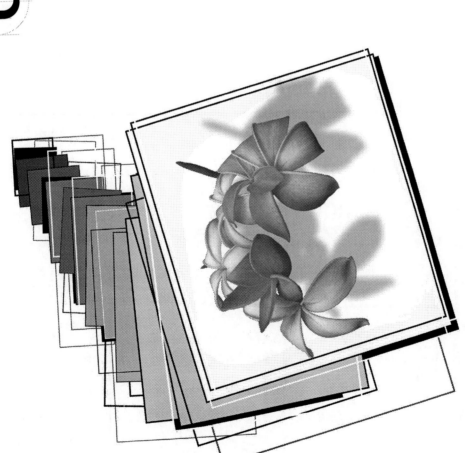

Ecology
Section 5

	LEVELS OF ORGANIZATION	OBTAINING ENERGY IN AN ECOSYSTEM	THE FLOW OF ENERGY IN AN ECOSYSTEM	ECOSYSTEM RELATIONSHIPS	CHEMICAL CYCLES
5	An individual of a species	An organism that uses sunlight or chemicals to make organic compounds from inorganic compounds	A pathway beginning with producers where energy is transferred from one organism to the next as organisms are consumed	The physical place or area where an organism lives	The chemical cycle that involves evaporation, transpiration, condensation, and precipitation
10	A group of individuals of the same species living in the same area at the same time	An organism that must consume other plants or animals to obtain energy	The interconnected feeding relationships among organisms in an ecosystem	The biological and physical role of an organism in its environment	The chemical cycle that involves photosynthesis, respiration, decomposition, and combustion
15	Different groups of species interacting and living in the same area	An organism that gains energy by eating only plants	An organism's feeding level in an ecosystem based on its nutritional source	Interaction in which one organism kills and consumes another organism	The chemical cycle that involves nitrogen fixation and denitrification

	The chemical cycle that involves a chemical that is released gradually from rock and is needed to make DNA and RNA	The process by which chemicals, elements, and matter are cycled between organisms and parts of the biosphere	A chemical that is scarce in an ecosystem but is needed by organisms; it controls the rate of growth among organisms based on its availability.
	A relationship in which two or more organisms live together and both organisms benefit	A relationship in which one organism lives in or on another organism and harms it by taking nutrients from it	A relationship in which one organism benefits from another organism and that organism is neither harmed nor helped
	A diagram that demonstrates the amount of energy in each trophic level	A diagram that demonstrates the number of individual organisms in each trophic level	A diagram that demonstrates the amount of living tissue in each trophic level
	An organism that gains energy by eating only animals	An organism that gains energy by eating both plants and animals	An organism that gains energy by breaking down dead plants and animals
	All of the living species in an area interacting with the nonliving environment	An area on land that shares a similar climate, soil condition, and dominant type of vegetation	The broadest level of organization that extends from Earth's atmosphere to below the ocean floor
	20	**25**	**BONUS**

NOTES

©2004 Walch Publishing

Biology Jeopardy 13

Ecology
Section 6

TERRESTRIAL BIOMES	AQUATIC ECOSYSTEMS	POPULATION CHANGE	ECOSYSTEM CHANGE I	ECOSYSTEM CHANGE II
5 A biome characterized by a warm climate and large amounts of rain and sunlight	An area where freshwater rivers meet the salt water of seas or oceans	The movement of individuals into a population	A process that occurs when a new community develops in an area that has not previously supported life and has no soil	The process of adding excess waste to the biosphere
10 A biome that receives less than 25 cm of rainfall each year, has warm temperatures during the day and cold temperatures at night, and has nutrient-poor soil	An area of land that is covered by fresh or salty water at least part of the year	The movement of individuals out of a population	A process of replacing a community in an area where a community has been destroyed but the soil remains	Precipitation that has a lower pH than normal precipitation because it contains nitric acid and sulfuric acid
15 A biome characterized by trees that lose their leaves in the fall, nutrient-rich soil, and year-round precipitation	The upper portion of the ocean that is penetrated by sunlight	A model of population growth in which there is a larger number of reproducing individuals in each subsequent generation	The first organism to populate an area that has no soil	A species whose population is decreasing so quickly that it is at risk of disappearing from Earth if the trend continues

20	A biome characterized by coniferous trees, long, cold winters, and summers that allow the ground to thaw	An area along the ocean shoreline that is covered by water at high tide and left uncovered by water at low tide	A model of population growth in which a population size increases rapidly and then levels off at or near a carrying capacity	A process in dry climates by which agricultural land is turned into desert	The disappearance of a species from its geographic range
25	A biome characterized by low growing vegetation and a layer of permafrost in the subsoil	These are found in warm, clear, shallow waters and are made of calcium carbonate exoskeletons of small animals.	A limiting factor that affects a population that has reached a certain density	A process of destroying forests	A trend of increasing temperatures on Earth
BONUS	A biome that receives too little rain to support trees and is dominated by grass species growing in nutrient-rich soil	An area found along the bottom of the ocean	A limiting factor that affects a population regardless of the population density	An area of the atmosphere that protects Earth from ultraviolet radiation but appears to be getting thinner	The range of biological species found on Earth
NOTES					

©2004 Walch Publishing

UNIT 3

Cell Biology

Cell Biology
Section 7

	Cell Discoveries	Cell Structures I	Cell Structures II	Cell Structures III	Cellular Organization
5	He claimed that cork tissue looked like tiny cells in a monastery when he observed the tissue through a microscope.	The organelle that controls most of the processes in the cell and contains the cell's genetic material	An organelle used to convert sunlight energy into food molecules during the process of photosynthesis	The membrane that surrounds the nucleus	The internal structures in a cell that perform a specific function
10	Using a microscope, he was the first person to observe small living organisms in water.	The jellylike material inside the plasma membrane in which cell structures are embedded	Known as the "powerhouse of the cell," this organelle converts food molecules into energy that is usable by cells.	The condensed genetic material that is made of DNA wrapped around proteins	A group of similar cells that work together to perform a task
15	This German botanist said all plants are made of cells.	The flexible boundary between a cell and its external environment	A sac surrounded by a membrane where fluid, proteins, carbohydrates, salts, and sometimes waste products are stored	The site inside the nucleus where ribosomes are made	The groups of tissues that function together to perform a task

Biology Jeopardy

©2004 Walch Publishing

20	This German scientist said all animals are made of cells.	The relatively inflexible layer found outside the cell membrane in some organisms	Small structures that store enzymes used to break down food particles, worn out parts of the cell, and invading bacteria and viruses	The hollow tubes of protein that are used to help a cell retain its shape and serve as tracks for organelles to move on inside the cell	A group of organs that work together to perform a function
25	This German physician stated that all cells come from preexisting cells.	These small structures made of RNA and proteins are the sites where proteins are assembled.	A folded membrane inside the cell that is covered with ribosomes	The solid fibers of protein that help a cell move and keep its shape	A type of cell that has a cell membrane and cytoplasm but does not have a nucleus
BONUS	This American proposed that some organelles in eukaryotic cells today were once free-living unicellular organisms.	Stacks of membranes where lipids and carbohydrates are added to proteins	A scaffold inside the cell made of microtubules and microfilaments	The folded membranes in which lipids are made inside the cell	A type of cell that contains a nucleus
NOTES					

Cell Biology
Section 8

	THE CELL MEMBRANE I	THE CELL MEMBRANE II	PHOTOSYNTHESIS AND ENERGY CARRIERS	PHOTOSYNTHESIS I	PHOTOSYNTHESIS II
5	The process by which molecules move from an area of high concentration to an area of lower concentration	The passive transport of molecules across a membrane using transport proteins	It contains three phosphate groups and is one of the main chemical compounds used to store energy in cells.	The energy source in photosynthesis	The location where the light-dependent reactions occur
10	The property of biological membranes that allows some molecules to pass through while keeping other molecules out	The process requiring energy by which materials are moved across a membrane against the concentration gradient	The number of phosphate groups in adenosine diphosphate or ADP	The gas needed for photosynthesis to occur	The stack of thylakoids inside a chloroplast
15	The diffusion of water across a selectively permeable membrane	The form of active transport in which cells surround and engulf materials	The name of the five-carbon sugar found in both ATP and ADP	The gas produced as a result of photosynthesis	A step in the light-dependent reaction in which light is absorbed by pigments

©2004 Walch Publishing

| 20 | A solution containing roughly the same concentration of dissolved particles or solutes as are contained inside a cell | The form of active transport in which cells expel large amounts of material | The protein in which ADP is converted into ATP when H⁺ ions pass through it | Six carbon compounds produced during photosynthesis | A step in the light-dependent reaction in which oxygen is separated from water |

| 25 | A solution containing a lower concentration of dissolved particles or solutes than are contained inside a cell | The osmotic pressure inside a plant cell causing cells to be rigid | The name of the process that converts ADP into ATP | A primary pigment used by plants to absorb mostly violet and red wavelengths of light | A step in the light-dependent reaction in which high energy electrons are passed from one molecule to another |

| BONUS | A solution containing a higher concentration of dissolved particles or solutes than are contained inside a cell | The loss of osmotic pressure that causes a cell membrane to pull away from a cell wall | The energy carrier made during photosystem I when high energy electrons are picked up by NADP⁺ | An accessory pigment in plants used to absorb mostly blue and red wavelengths of light | A step in the light-dependent reaction in which high-energy electrons are used to convert NADP⁺ to NADPH |

NOTES

Cell Biology
Section 9

	PHOTOSYNTHESIS III	CELLULAR RESPIRATION I	CELLULAR RESPIRATION II	CELLULAR RESPIRATION III	CELL REPRODUCTION I
5	A step in photosynthesis that does not require light	The type of cellular respiration that requires oxygen	The net gain of ATP in glycolysis	An enzyme that combines with pyruvic acid in aerobic respiration	The genetic material that is divided during cell reproduction
10	The location in the cell where the Calvin cycle occurs	The type of cellular respiration that does not require oxygen	A three-carbon molecule produced during glycolysis	A product made when a two-carbon molecule combines with Coenzyme A	The proteins found in eukaryotic cells that DNA is wrapped around
15	The five-carbon compound used in the Calvin cycle	The organelles in which the final stages of cellular respiration occur	The electron carrier produced during glycolysis	A four-carbon compound that combines with acetyl-CoA to begin the Krebs cycle	The rod-shaped structure found in eukaryotic cells during cell division that is made of DNA and proteins

20	Two three-carbon molecules that are made when a CO_2 molecule is added to a five-carbon compound	The first stage of cellular respiration	A type of anaerobic respiration that produces two ethyl alcohol molecules: carbon dioxide and NAD^+	A molecule made when a four-carbon compound joins with acetyl-CoA	One of two identical parts of a duplicated chromosome
25	Three-carbon compounds that result when energy from ATP and NADPH is used to convert PGA into a higher energy form	The first molecule to be broken down in cellular respiration	A type of anaerobic respiration that occurs in human muscle cells during strenuous exercise	The stage of aerobic respiration that uses high-energy electrons to carry H^+ ions across the inner membrane of mitochondria	An area where two sister chromatids are held together in a chromosome
BONUS	The pathway that allows plants to fix carbon using four-carbon compounds	The location where glycolysis occurs	The stage of aerobic respiration that involves a number of steps involving pyruvic acid and produces carbon dioxide	An enzyme used to convert ADP to ATP	A series of events that a cell goes through as it grows and divides
NOTES					

Cell Biology
Section 10

3

	CELL REPRODUCTION II	CELL REPRODUCTION III	CELL REPRODUCTION IV	CELL REPRODUCTION V	CELL REPRODUCTION VI
5	The period of cell growth and development between cell divisions	The process of nuclear division in eukaryotic cells	An area where the cell membrane pinches together during cell division	Cell division that results in four gametes, each having half the normal number of chromosomes for a species	The process by which homologous chromosomes exchange equal segments of genetic material during meiosis
10	The period during interphase when the DNA is copied or synthesized	The second phase of mitosis during which chromosomes line up across the center of the cell	The membrane that forms between two new plant cells during cell division	A pair of chromosomes that are identical in length, centromere location, and gene locations; one is inherited from each parent.	The phase of meiosis during which crossing-over occurs
15	The period during interphase when the cell prepares for and makes cell structures needed for cell division	The first phase of mitosis during which chromatin condenses into chromosomes	Two identical cells formed from one original cell during cell division	A cell containing two sets of chromosomes in which each set is a homologous pair	The phase in meiosis I during which homologous chromosomes separate and move to opposite ends of the cell

20	25	BONUS	NOTES
The first stage of interphase during which the cell grows	The third phase of mitosis during which chromosomes separate and move to opposite ends of the cell	Proteins that fluctuate in the cell and help to regulate the cell cycle	A cell containing only one set of chromosomes	The phase in meiosis II during which chromosomes line up at the middle of the cell
Two small structures made of microtubules that are located in the centrosome during cell division in animal cells	The fourth phase of mitosis during which the nuclear envelope reforms around each new set of chromosomes	A disorder that results from the uncontrolled growth of cells	A cell division in males that results in the production of sperm cells	The phase in meiosis II during which chromatids separate and move to opposite ends of the cell
Microtubules that extend across the cell during cell division	The division of cytoplasm into two new cells	A mass of cells that results from abnormal or uncontrolled cell division	A cell division in females that results in the production of a mature egg	The phase in meiosis II during which the nuclear envelope reforms around genetic material in four new haploid cells

©2004 Walch Publishing

Biology Jeopardy 25

UNIT 4

Genetics

Genetics
Section 11

	MENDELIAN GENETICS I	MENDELIAN GENETICS II	MENDELIAN GENETICS III	GENETIC CROSSES I	GENETIC CROSSES II
5	An Austrian monk who is considered the father of modern genetics	The process in which pollen of one plant lands on the stigma of the same plant	An allele that is always expressed when it is present in an organism	Used to determine the likelihood that an event will occur	A cross involving only one trait
10	Organisms grown in Mendel's garden for his experiments	The process in which pollen of one plant lands on the stigma of a different plant	An allele that will only be expressed when a dominant allele is not present in an organism	A diagram used by biologists to help determine the outcome of a genetic cross	A cross involving two traits
15	The passage of characteristics from parent to offspring	Plants that have only one form of an allele for a trait	A trait that is recessive to tall plant height	An organism that has two of the same alleles for a trait	A cross of an organism with unknown genotype and a homozygous recessive organism to determine the unknown genotype

4

28 *Biology Jeopardy* ©2004 Walch Publishing

20					
The study of heredity	The offspring of the parent generation	A trait that is dominant to yellow pea color	An organism that has two different alleles for a trait	The phenotypic ratio of a cross between a homozygous tall (TT) plant and a homozygous short (tt) plant	
25					
The characteristics Mendel observed in contrasting pairs in his pea plants	A product of meiosis	A trait that is recessive to round pea shape	The physical characteristic expressed by an organism for a trait	The genotypic ratio of a cross between a homozygous tall (TT) plant and a homozygous short (tt) plant	
BONUS					
A different form of a gene	The principle stating that alleles separate from one another during gamete formation	The principle stating that an allele for a trait may be passed on, independent of another allele during gamete formation	The genes possessed by an organism for a trait	The phenotypic ratio of a cross of a plant homozygous for round and yellow seeds (RRYY) and a plant homozygous for wrinkled and green seeds (rryy)	NOTES

Genetics
Section 12

4

NUCLEIC ACIDS I	NUCLEIC ACIDS II	NUCLEIC ACIDS III	NUCLEIC ACIDS IV	NUCLEIC ACIDS V
The structure that stores and transmits genetic information in cells	The five-carbon sugar found in DNA	A term used to describe the shape of DNA	The five-carbon sugar found in ribonucleic acid (RNA)	The main enzyme used during transcription; it separates the DNA strands and binds mRNA to the DNA.
A process described by Frederick Griffith when he observed harmless bacteria take on characteristics of disease-causing bacteria	A double-ringed nitrogenous base	Weak bonds that hold nitrogenous bases together in base pairing	A nitrogen base found in RNA but not in DNA	The process known as protein synthesis in which the mRNA code is used to build a polypeptide chain
A scientist who, with other researchers, determined that DNA is the molecule that stores and transmits genetic information	A single-ringed nitrogenous base	The process by which DNA makes a copy of itself	A type of RNA that copies the genetic code in the nucleus and takes it to the ribosomes	The location where translation occurs

5

10

15

			NOTES
A sequence of three mRNA nucleotides	A type of RNA that combines with proteins to make ribosomes	The location where the DNA separates during replication	
A sequence of three tRNA nucleotides that bind with a codon during translation	A type of RNA that brings amino acids to the ribosomes	An enzyme that breaks the hydrogen bonds between nitrogenous bases in DNA before replication	
Monomers carried by the tRNA that are used to make a polypeptide	The process by which DNA is separated and mRNA makes a copy of the DNA code	An enzyme that binds to each separated strand of DNA and joins nucleotides to build a complementary strand of DNA	

20	A nitrogenous base that pairs with thymine	Two scientists who used radioactive markers to reinforce that DNA is the molecule that stores and transmits genetic information
25	A nitrogenous base that pairs with guanine	A scientist who studied the structure of DNA using X-ray diffraction
BONUS	A nitrogenous base found in DNA but not in RNA	Two scientists who published a paper in 1953 suggesting the correct structure of DNA

Genetics
Section 13

	GENE EXPRESSION	GENETIC ERRORS	INHERITANCE PATTERNS	HUMAN INHERITANCE I	HUMAN INHERITANCE II
5	A segment of DNA to which RNA polymerase binds to begin transcription	A change in DNA that affects the sequence of nucleotides	The traits controlled by a gene found on the X or Y chromosome	A diagram that shows relationships among family members and can be used to trace inheritance of a trait over several generations	The sex chromosome present in males that is not present in females
10	A DNA segment that does not code for a protein and is removed before transcription	A gene mutation caused by the substitution of one nucleotide	An inheritance pattern in which three or more alleles exist for the same gene	An individual who has an allele for a trait but does not express the trait	A condition in which there is only one chromosome of a chromosome pair in a cell as a result of nondisjunction during meiosis
15	A DNA segment that does code for a protein	A mutation caused by the insertion or deletion of one nucleotide	An inheritance pattern in which two alleles for a gene contribute to the phenotype of an organism	Chromosomes that determine the gender of an individual	A condition in which there are three of the same chromosome in a cell due to nondisjunction during meiosis

A condition in which there are three number 21 chromosomes in the cells of an individual	A condition in which a female has only one X chromosome as a result of nondisjunction during meiosis	A condition in which a male has two X chromosomes and one Y chromosome as a result of nondisjunction during meiosis	
All of the chromosomes that do not determine the gender of an individual	A photograph of chromosomes grouped by homologous pairs	An inactivated X chromosome that becomes a dense object inside the nucleus	
An inheritance pattern in which both alleles contribute to the phenotype of an organism and one allele is not dominant	The traits controlled by alleles in two or more different genes	When homologous chromosomes fail to separate during anaphase I of meiosis	
A chromosomal mutation caused by the loss of part or all of a chromosome	A chromosomal mutation caused by reversing the direction of a piece of a chromosome	A chromosomal mutation that occurs when a piece of a chromosome breaks off and attaches to a nonhomologous chromosome	
Normally found in prokaryotes and phages, this group of genes works together.	A protein that inhibits or suppresses the expression of a gene	A DNA sequence found in the operon to which a protein binds to inhibit transcription	
20	**25**	**BONUS**	**NOTES**

©2004 Walch Publishing

Biology Jeopardy 33

Genetics
Section 14

4

HUMAN INHERITANCE III	HUMAN INHERITANCE IV	DNA RESEARCH I	DNA RESEARCH II	DNA RESEARCH III
A disorder caused by a recessive allele on the X chromosome that causes individuals to have difficulty distinguishing colors	A genetic disorder caused by a recessive allele on an autosome in which affected individuals lack pigment in their skin, eyes, and hair	One of the first insects used for genetic research	The process of manipulating and making changes in the DNA of an organism	A short, single strand of DNA used in PCR to give DNA polymerase a starting point to begin replication
A disorder caused by a recessive allele on the X chromosome that causes individuals to not produce a protein needed in blood clotting	A genetic disorder caused by a gene mutation that results in a misshapen hemoglobin protein, changing the shape of red blood cells	A scientist whose research in the early 1900s showed that genes were inherited together in linkage groups	DNA combined from two or more organisms	The process of making a genetically identical individual
A genetic disorder caused by a recessive allele on the X chromosome that causes skeletal muscles to weaken and atrophy	A genetic disorder that often appears in middle age as a result of a dominant allele on an autosome that causes a loss of muscle control and mental deterioration	A diagram developed by Alfred Sturtevant showing the distance between genes based on recombination rates	A circular piece of DNA found in bacteria	The name given to the first cloned sheep

5

10

15

Biology Jeopardy

©2004 Walch Publishing

A disorder caused by a recessive allele on an autosome that causes individuals to make misshapen chloride channel proteins	An inheritance pattern responsible for ABO blood groups	Begun in 1990, this international project was designed to sequence all of human DNA.	An enzyme that cuts DNA at specific nucleotide sequences	An organism containing recombinant DNA
A genetic disorder caused by a recessive allele on an autosome that causes lipid accumulation in the central nervous system	First discovered in rhesus monkeys, this is found on red blood cells.	The process of separating fragments of DNA using gel and an electric current	A single strand of DNA that remains after DNA has been cut by a restriction enzyme	The banding pattern of DNA fragments that are specific to an individual
A disorder caused by a recessive allele on an autosome that causes individuals to not make an enzyme needed to break down phenylalanine	Medical advice given to couples concerned that their offspring may be at risk for a genetic disorder	The process of replacing a faulty gene with a gene that functions properly	A technique used to make multiple copies of DNA segments	Two scientists who announced in June 2000 that the human genome sequence was essentially completed

20

25

BONUS

NOTES

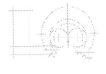

UNIT 5

Evolution

Evolution
Section 15

	EARTH'S HISTORY I	EARTH'S HISTORY II	ORIGIN OF LIFE	EVOLUTION I	EVOLUTION II
5	The preserved remnant or trace of an organism from a past geologic age	A Soviet scientist who hypothesized the composition of Earth's early atmosphere	The idea that life arises from nonliving matter	The individual who published *On the Origin of Species* in 1859	An economist whose views on human population growth influenced Darwin
10	A scientist who studies fossils	American scientists whose experiment in 1953 was set up to simulate conditions on primitive Earth	The principle that life arises only from other life	The sailing ship on which Darwin traveled for five years	A method used by humans to choose and breed individuals of a species with the most desirable traits
15	A technique used to determine the age of a fossil by comparing the location of a fossil with other fossils, especially index fossils	Cell-like structures with a membrane of protein molecules made from organic compounds	A scientist who set up a controlled experiment in the 1600s that proved maggots develop from fly eggs and not from nonliving matter	The location off the coast of South America where Darwin collected finches	The ability of an organism to survive and reproduce in an environment and thereby pass on their genes to their offspring

©2004 Walch Publishing

20	A technique that uses radioactive isotopes to determine the age of material	Believed to be the first nucleic acids to have formed on Earth	A scientist who set up a controlled experiment in the 1700s to show that microorganisms come from other microorganisms	An individual whose ideas on evolution were similar to Darwin's; his essay on these ideas prompted Darwin to publish *On the Origin of Species*.	An inherited trait that increases an organism's chance of survival in a particular environment
25	The time required for half of a radioactive isotope to decay	The protective layer in the upper atmosphere made from oxygen produced by early photosynthetic organisms	A scientist whose experiment in the 1800s proved that microorganisms develop from other microorganisms and not from a "vital force"	A scientist who hypothesized that acquired characteristics are passed from parents to offspring	The process by which individuals with variations that make them best-suited to their environment reproduce more than individuals without those variations
BONUS	The first and longest era in Earth's history	The theory stating that eukaryotic cells developed as a result of mutually beneficial relationships between prokaryotic cells	The theory stating that organisms change over time	A scientist whose *Principles of Geology* influenced Darwin's thinking about the age of Earth and its history	Individuals do not evolve, but these do.

NOTES

©2004 Walch Publishing

Biology Jeopardy

Evolution
Section 16

	EVOLUTION III	EVOLUTION IV	EVOLUTION V	EVOLUTION VI	PRIMATES
5	Structures that differ in the adult form of different organisms but are derived from the same embryonic tissues	A condition in which the allelic frequency stays the same in a population	A form of natural selection in which individuals with an average form of a trait are not selected for and individuals at either extreme are selected for	A process by which many organisms become extinct at the same time	A digit that is oriented opposite of the other fingers and allows a primate to grasp objects
10	Structures that have a similar function in adults of different organisms but are derived from different embryonic tissues	The principle developed by two mathematicians that shows what must occur in a population to maintain genetic equilibrium	The formation of a new species	A pattern of evolution in which two species interact closely and change in response to each other over time	The ability to walk upright on two legs
15	Structures that served a purpose in an ancestral organism but do not have a useful function in the modern organism	The random change that occurs in a gene pool of a small population due to chance	An isolating mechanism that occurs when populations cannot interbreed and produce fertile offspring	A pattern of evolution in which one species evolves into several different species	A group of primarily nocturnal primates that includes lemurs and tarsiers

©2004 Walch Publishing

Darwin's theory that current species descended from a common ancestor with changes over time	The process by which the gene pool of a small emigrating population can change the allelic frequency of a new population	An isolating mechanism that occurs when populations are physically separated by a barrier, such as a river or mountain range	A pattern of evolution in which different species evolve to have similar structures as a result of living in similar environments	A group of primates that includes monkeys, apes, and humans
All of the genes in a given population	A form of natural selection in which individuals with a more extreme form of a trait are selected over individuals with an average form of a trait	An isolating mechanism that occurs when populations are separated by behaviors, such as variations in courtship rituals	A pattern of evolution in which stable periods in evolutionary history are interrupted by brief periods of rapid change	An appendage used by New World monkeys that is like a fifth hand to grasp objects
The number of times an allele occurs in a population compared to the total number of times other alleles occur	A form of natural selection in which individuals with an average form of a trait are selected and have the highest fitness	An isolating mechanism that occurs when populations are separated because they reproduce at different times	A pattern of evolution in which there is a slow and steady change in a species over time	A group of primates that have a large cranial capacity for their body size that includes chimpanzees, orangutans, gibbons, and gorillas

20

25

BONUS

NOTES

©2004 Walch Publishing

Biology Jeopardy 41

Evolution
Section 17

	HUMAN EVOLUTION	CLASSIFICATION I	CLASSIFICATION II	CLASSIFICATION III	CLASSIFICATION IV
					In a six-kingdom system, the kingdom that includes the multicellular, heterotrophic organisms such as sponges, insects, and primates
5	A group consisting of humans and their most recent ancestors	The branch of biology that names organisms and classifies them according to their physical characteristics and evolutionary history	A group of related classes	A diagram showing evolutionary relationships based on morphological differences, fossil records, embryology, and chromosomal and biochemical comparisons	
					In a six-kingdom system, the kingdom that includes multicellular, autotrophic organisms such as mosses and flowering plants
10	A group of early hominids from the genus *Australopithecus*	The system of giving an organism a two-part scientific name in Latin	A group of related orders	A feature that appears only within recent parts of a lineage and is absent in older parts of the lineage	
					In a six-kingdom system, the kingdom that includes unicellular and multicellular heterotrophs such as mushrooms, yeast, and molds
15	The name given to the female fossil of *Australopithecus afarensis* discovered by Donald Johanson	A Swedish scientist who developed the system of naming and classifying organisms into groups during the 1700s	A group of related families	A diagram that shows evolutionary relationships based on derived characteristics	

An extinct member of the genus *Homo* known as a "handy man" because stone and bone tools were found near the fossil remains	The system used to identify a species based on a series of opposing statements	A group of related genera	A method of classification that uses derived characteristics to establish an organism's most recent ancestor	In a six-kingdom system, the kingdom that includes unicellular and multicellular autotrophs and heterotrophs such as amoeba and algae
20				
An extinct member of the genus *Homo* known as the "upright walker" who may have been the first in the genus to migrate out of Africa	The largest taxonomic level developed by Linnaeus	A group of related species	A classification level now used that is higher than a kingdom	In a three-domain system, the domain that includes prokaryotic, unicellular organisms that have peptidoglycan in their cell walls
25				
A group of *Homo sapiens* who appeared in Europe and are known for their cave paintings in France and evidence of elaborate burials of their dead	The next level under kingdom in the plant kingdom	A group of organisms of one type that can breed and produce fertile offspring	In a three-domain system, the domain that includes eukaryotic organisms such as protists, fungi, plants, and animals	In a three-domain system, the domain that includes prokaryotic, unicellular organisms that do not have peptidoglycan in their cell walls
BONUS				

NOTES

UNIT 6

Microorganisms and Fungi

Microorganisms and Fungi
Section 18

Bacteria Classification	Characteristics of Bacteria I	Characteristics of Bacteria II	Characteristics of Bacteria III	Viruses I
5 The kingdom or domain of bacteria that are characterized by the lack of peptidoglycan in the cell wall and unusual lipids in the cell membrane	Bacteria that are round or spherical in shape	Bacteria, like cyanobacteria, that use sunlight to make energy	Asexual reproduction in which bacteria copy their genetic information and divide in half	The outer protein coat of a virus
10 The kingdom or domain of bacteria that are characterized by having peptidoglycan in the cell wall	Bacteria that are rod-shaped	Bacteria that obtain energy by feeding on dead and decaying matter	Sexual reproduction in bacteria in which DNA is exchanged through joined pili	A virus that infects bacteria
15 When exposed to Gram stain, these bacteria turn violet in color.	Bacteria that are spiral-shaped	Bacteria that make energy by using chemicals instead of light	A structure that allows a bacteria to go dormant; it contains cytoplasm and DNA encased in a tough coating.	A process by which a bacteriophage enters a host cell, makes copies of itself, and then destroys the host cell when new viruses are released

6

©2004 Walch Publishing

			A process by which the genetic code of a bacteriophage is replicated with the host cell's DNA for a period of time without destroying the cell	
			A piece of a bacteriophage's genetic code that is embedded in the host's DNA during the lysogenic cycle	
			Found on the envelope of a virus, these projections of proteins, containing chains of sugar, help viruses attach to host cells.	
		A type of chemical that inhibits the growth of bacteria		
		A disease-causing bacteria		
		The process bacteria use to convert nitrogen gas into ammonia		
	Bacteria that require oxygen to live			
	Bacteria that cannot survive in the presence of oxygen			
	Bacteria that can live with or without oxygen			
	Spherical-shaped bacteria that live in long chains			
	Spherical-shaped bacteria that live in clusters			
	A structure some bacteria use for movement			
20 — When exposed to Gram stain, these bacteria turn pink in color.				
25 — A group of anaerobic bacteria that make energy by converting H_2 and CO_2 into methane gas				
BONUS — A group of bacteria that live in extremely acidic and hot environments, such as hydrothermal vents				
NOTES				

©2004 Walch Publishing

Biology Jeopardy 47

Microorganisms and Fungi
Section 19

	VIRUSES II	PROTIST PHYLA I	PROTIST PHYLA II	PROTIST CHARACTERISTICS I	PROTIST CHARACTERISTICS II
5	A viral capsid shape made of a geometric pattern of 20 triangular surfaces	A group of heterotrophic protists that have one or more flagella; members of this group cause African sleeping sickness, giardia, and leishmaniasis.	A group containing the green algae; their chloroplasts have both chlorophyll a and b and their cell walls are made of cellulose.	An extension of cytoplasm used for feeding and movement in some organisms	A structure used by algae to attach to a substrate
10	Viruses that contain RNA and reverse transcriptase	A group of heterotrophic protists that use cilia for locomotion and feeding; members of this group are in the genus *Paramecium*.	A group containing the red algae; they contain chlorophyll a and phycobilins.	An organelle used to collect and expel water in protists	A gellike compound extracted from algae used to grow bacteria and other microorganisms in culture
15	A virus known to cause acquired immunodeficiency syndrome (AIDS)	Members of this group contain photosynthetic pigments, lack a cell wall, and move using two flagella; some species are in the genus *Euglena*.	A group containing the brown algae; they contain both chlorophyll a and c and fucoxanthin.	Food swept into the oral groove of a ciliate is forced into a vacuole at this site.	The light produced by organisms such as dinoflagellates as a result of a chemical reaction

6

48 Biology Jeopardy ©2004 Walch Publishing

20				
A type of virus known to cause cancer in animals	A group containing protists that do not move on their own and reproduce using sporozoites; many species are parasitic.	A group containing the diatoms that have silicon in their cell walls	A collection of pigments that helps protists and some invertebrates detect light	A structure found in acellular slime molds that contains many nuclei

25				
Small strands of RNA nucleotides, lacking a capsid, that are known to infect plants	A group of heterotrophic protists that move and feed using pseudopodia; some are encased in a shell (radiolarians) and some are not (amoebas).	A group containing yellow-brown algae that store much of their extra energy as oil	A structure used to exchange genetic information during conjugation in *Paramecium*	The sexual reproduction in algae and plants that switches between a haploid phase and a diploid phase

BONUS			
Infectious proteins that cause proteins to become abnormally shaped; linked with transmissible spongiform encephalopathies	In this group, about half of the species are photosynthetic and half are not; they have two flagella and cell walls made of cellulose.	A group containing the cellular slime molds	A tough but flexible covering made of proteins found around ciliates and euglenoids	The common name given to funguslike organisms in phylum Oomycota that were responsible for the Potato Famine

NOTES

Microorganisms and Fungi
Section 20

	PROTIST CHARACTERISTICS III	FUNGI I	FUNGI II	FUNGI III	POTPOURRI
5	The membrane-bound sacs of digestive enzymes in which food is broken down	A phylum of fungi, known as sac fungi, that develop haploid spores in an ascus	A polysaccharide containing nitrogen that is found in the cell walls of fungi	A unicellular sac fungus used to make bread rise and to ferment grains and grapes	Solutions containing a weakened form of a pathogen, such as a virus, bacteria, or toxin, that causes an immune response in the body
10	A flattened leaflike portion of a seaweed that collects light for photosynthesis	A phylum of fungi that includes the black bread mold and uses zygospore during its life cycle	Tiny filaments found in multicellular fungi	Hyphae that grow into and anchor fungi, such as black bread mold, to a substrate	Small plant and animal organisms that swim or float in aquatic environments
15	A defense structure in paramecium that releases stiff projections from the cell	A phylum of fungi, known as the club fungi, that use spore-bearing basidia during their life cycle	The cross wall found in the hyphae of some fungi	Hyphae that grow across a substrate in a fungus such as black bread mold	An antibiotic derived from fungi used to inhibit cell wall production in bacteria

20	25	BONUS
The sticky polysaccharide found in some red algae that is used in ice cream, cosmetics, and some cheeses	The form of sexual reproduction used by paramecium to exchange genetic material between individuals	The multinucleated mass of cytoplasm found in acellular or plasmodial slime molds
A phylum of fungi known as the imperfect fungi because a sexual stage of their life cycle has not been identified	The symbiotic relationship between a fungus and a photosynthetic organism such as a green algae or cyanobacteria	The symbiotic relationship between a fungus and plant roots
A thick tangled mat of hyphae	In some fungi, this saclike structure contains spores and sits on top of a sporangiophore	A structure containing gametes that forms at the end of each mating type of hyphae during sexual reproduction
Structures found in sac fungi in which ascospores develop during sexual reproduction	Structures found in sac fungi in which conidia develop during asexual reproduction	Club-shaped structures containing spores found in the club fungi
An infection caused by a fungus that results in itching and splitting of the top layer of skin	A type of colonial green algae that form a sphere and live in freshwater	Food poisoning caused by anaerobic bacteria that release toxins into canned foods that have not been completely sterilized

NOTES

Biology Jeopardy

Unit 7

Plants

Plants
Section 21

Plant Phyla I	Plant Phyla II	Mosses and Their Relatives	Ferns and Their Relatives	Plant Cells and Tissues I
The phylum of plants that lack vascular tissue and produce spores; includes mosses like Sphagnum	The phylum of plants that have vascular tissue and produce spores; includes the ferns	The structure on top of a stalk in a moss sporophyte that contains spores	The mature leaf of a fern	The type of tissue that forms the outer covering on plants
The phylum of plants that lack vascular tissue and produce spores; includes the liverworts	The phylum of plants that have vascular tissue and produce seeds in cones; includes the conifers such as pines, spruces, firs, and cedars	The rootlike structure that attaches moss gametophytes to the soil	An underground stem that anchors a fern to the soil	A single layer of dermal tissue covering the outer surface of plants
The phylum of plants that lack vascular tissue and produce spores; includes the hornworts	The phylum of plants that have vascular tissue and produce seeds in cones; has only one living species today, *Ginkgo biloba*	The structure in the male gametophyte that produces sperm cells	The structure on the underside of a frond that contains spores	A thick, waxy layer on the outer surface of cells that prevents water loss

5
10
15

20	The phylum of plants that have vascular tissue and produce spores in a strobilus; includes the club mosses	The phylum of plants that have vascular tissue and produce seeds in cones; includes the cycads	The structure in the female gametophyte that produces an egg cell	A cluster of sporangia	A type of tissue that provides new growth when cells divide by mitosis
25	The phylum of plants that have vascular tissue and produce spores at the end of short branches; includes the whisk ferns	The phylum of plants that have vascular tissue and produce seeds in cones; includes three different-looking genera: *Ephedra*, *Welwitschia*, and *Gnetum*	The green filaments that grow from a spore when it germinates	The thin, heart-shaped gametophyte that forms when a spore germinates	The region of growth that allows plants to grow in length at the root tips and at the end of stems
BONUS	The phylum of plants that have vascular tissue, produce spores and have silica in their cell walls; includes the scouring rushes	The phylum of plants that have vascular tissue and produce seeds in fruit; includes flowering plants such as roses, apples, corn, and grass	A cup-shaped structure that contains haploid cells used in asexual reproduction in liverworts	A cone made of leaves bearing sporangia found in liverworts	The region of growth that allows plants to grow in width

NOTES

Plants
Section 22

Plant Cells and Tissues II	Plant Cells and Tissues III	Roots	Stems I	Stems II
A type of tissue that transports water and nutrients through a plant	A type of cell found in phloem that helps control the movement of sugars through a sieve-tube element	A layer of cells that covers the apical meristem and produces an oily lubricant as the root grows through the soil	An area that can develop into a new leaf or shoot	A layer outside a stem formed by the cork, cork cambium, and phloem
A type of vascular tissue that carries water up a plant from the roots to the leaves	The end wall of sieve-tube elements	The main root that grows long and thick	An area where a leaf joins the stem	The dark wood made of xylem cells at the center of a tree trunk
A type of vascular tissue that carries nutrients and carbohydrates down a plant from the leaves and stem to the roots	All of the tissue found between the dermal tissue and the vascular tissue	The roots that branch out and do not develop one main root	A section of the stem between nodes	The lighter wood made of xylem cells inside the vascular cambium

Row values: 5, 10, 15

			The bands of light and dark wood that can be seen when a tree trunk is cross-sectioned
			An underground stem that stores food, such as a potato
			An underground stem surrounded by thick leaves that store food, such as in a tulip
	The tissue at the center of a stem		
	The tissue that forms a ring around a stem and produces xylem to the inside and phloem to the outside of the stem		
	The tissue that produces cork as the outer covering of a stem		
	The tiny extensions out of the epidermis that create a large surface area for the root		
	A layer of ground tissue beneath the epidermis		
	A layer of cells outside the vascular tissue that is surrounded by the Casparian strip		
	Ground tissue cell types that have cube-shaped cells with large vacuoles and thin cell walls		
	Ground tissue cell types with thick, irregular-shaped cell walls that provide plant support and may be found in strands		
	Ground tissue cell types with thick and rigid cell walls that provide support and strength; these cells may die at maturity.		
20	A type of cell found in the xylem of all seed plants that has thick cell walls and tapered ends containing pits		
25	A type of cell found in the xylem of angiosperms that have either large holes or no cell wall on the ends of the cell		
BONUS	A type of cell found in phloem that has a nucleus and organelles pushed to the side of the cell so nutrients and carbohydrates can be transported		
NOTES			

Plants
Section 23

	LEAVES	FLOWERS	REPRODUCTION IN SEED PLANTS I	REPRODUCTION IN SEED PLANTS II	SEEDS AND FRUIT I
5	The flattened portion of a leaf where most photosynthesis occurs	The brightly colored structures that attract pollinators	The location where pollen grains are produced in gymnosperms	The sugar solution found in flowering plants that pollinators may feed on	The protective layer around a seed embryo
10	A stalk that connects the blade to the stem	The modified leaves that make the outer layer of a flower and protect the bud and flower as it develops	The location where ovules are produced in gymnosperms	A type of pollination most common in gymnosperms	A structure formed from a mature ovary wall in a flowering plant
15	A bundle of vascular tissue in a leaf	The male reproductive structure that includes the filament and anther	The type of cell that divides to produce pollen grains in a flowering plant	The nutrient-rich 3N tissue that forms from the cell created when a sperm cell fertilizes two polar nuclei	The seed leaf in a plant embryo

20	25	BONUS	NOTES
The column-shaped cells on the upper surface of a leaf where most photosynthesis occurs	The loosely packed cells surrounded by air spaces that allow for the exchange of oxygen and carbon dioxide in leaves	The opening on a leaf that is regulated in size by guard cells	
The female reproductive structure that makes the female gametophyte	The sticky area at the top of the style that helps trap pollen grains	The area at the base of the pistil where ovules are formed	
A type of cell that divides to produce eight cells in the ovule of a flowering plant	A structure that grows toward an ovule after a pollen grain reaches the stigma	The opening to the ovule located near the egg	
A type of fertilization unique to angiosperms in which the ovule is fertilized to form a zygote and the polar nuclei are fertilized to form endosperm	Asexual reproduction using nonreproductive parts of the plant such as leaves, stems, and roots	Asexual reproduction in which one plant is manually attached to a second plant's stem or roots	
An angiosperm seed with a single cotyledon	An angiosperm seed with two cotyledons	The stage when a seed begins to grow and develop	

Biology Jeopardy 59

Plants
Section 24

	SEEDS AND FRUIT II	PLANT HORMONES	PLANT RESPONSES I	PLANT RESPONSES II	POTPOURRI
5	The first part of a seed embryo to emerge—it develops into a root.	The hormone that stimulates cell elongation	A plant response to an environmental stimulus in a specific direction	The dropping of leaves, fruits, flowers, and stems after a layer of specialized cells form as a result of the release of ethylene gas	Plants that have been selected for and bred by humans to cultivate desirable traits
10	The portion of the embryo that develops into the stem between the radicle and cotyledons	The gas that causes fruit to ripen and promotes leaf abscission	A plant response to light	A plant response to the number of daylight and nighttime hours	Plants that are adapted to live in dry environments
15	The portion of the embryo that develops into the stem above the cotyledons	The hormones that cause stems to elongate and seeds to germinate	A plant response to gravity	The plant pigment that causes plants to respond to the number of daylight and nighttime hours	A plant that grows directly on top of another plant

20	25	BONUS	NOTES
In seeds, the period of time during which an embryo is alive but not growing	A hormone that works with auxin to promote cell elongation but also delays aging in leaves and causes dormant seeds to germinate	A plant response to touching an object	A plant that flowers when days are short and nights are longer than a critical length	An angiosperm that completes its life cycle in one growing season
Dry fruit that splits in half like a pea or peanut	A synthetic form of auxins used to kill weeds and defoliate plants	A plant response to chemicals	A plant that flowers when days are long and nights are shorter than a critical length	An angiosperm that requires two years to complete its life cycle
Fleshy fruit that has a stony layer around the seed, such as a coconut, peach, or cherry	A hormone that promotes dormancy in buds and seeds and helps plants respond to stressful conditions	A plant response to temperature	A plant that will flower regardless of the day length	An angiosperm that grows for more than two years

©2004 Walch Publishing

Biology Jeopardy

UNIT 8

Invertebrates

Invertebrates
Section 25

	CHARACTERISTICS OF THE ANIMALS I	CHARACTERISTICS OF THE ANIMALS II	CHARACTERISTICS OF THE ANIMALS III	CHARACTERISTICS OF THE ANIMALS IV	CHARACTERISTICS OF THE ANIMALS V
5	An animal lacking a backbone or vertebral column	A hollow ball of cells that develops after a zygote divides a number of times	The front or head end of an animal	A system in which circulatory fluid is not always contained in a network of blood vessels	The sessile organism that strains food from the surrounding water
10	An animal with a backbone or vertebral column	The inner layer of cells in a gastrula that develops into most of the respiratory system and lining of the digestive system	The back or tail end of an animal	A system in which blood is contained in a network of blood vessels	Digestion that occurs inside cells
15	An external skeleton that protects the soft tissues of an animal	The middle layer of cells in a gastrula that develops into the skeleton, muscles, and circulatory system	The top or upper side of an animal	An animal lacking a body cavity or a coelom	The union of gametes inside the female's body

20	25	BONUS
An internal skeleton that supports the body of an animal	An animal body that has repeated units each containing similar structures, such as those seen in earthworms	A skeleton consisting of a fluid-filled cavity surrounded by muscle that can be contracted to provide movement for an organism
The outer layer of cells in a gastrula that develops into the skin, hair, nails, and nervous system	An animal in which the blastopore develops into a mouth	An animal in which the blastopore develops into an anus
The bottom or lower side of an animal	Animals that have repeated body parts branching out from the center of the body	This term means "a condition of having two equal halves on either side of a central plane."
An animal with a body cavity partially lined with mesoderm and an endodermic gut suspended in fluid	An animal with a body cavity fully lined with mesoderm and an endodermic gut lined and suspended by mesoderm	A concentration of nerve tissue and sense organs at the front end of an animal
The union of gametes outside of the female's body	An organism that produces both male and female gametes	The free-living immature stage of an organism that looks different from the adult form

NOTES

Biology Jeopardy

Invertebrates
Section 26

INVERTEBRATE CLASSIFICATION I	INVERTEBRATE CLASSIFICATION II	SPONGES	CNIDARIANS AND CTENOPHORES I	CNIDARIANS AND CTENOPHORES II
The phylum that includes all sponges	The phylum that includes segmented worms such as the sandworms, earthworms, and leeches	A term for an organism that is attached to a surface and does not move from that location	The sessile form of a cnidarian that is cylindrical in shape with one end attached to the substrate and the mouth and tentacles extending up	The nervous system in cnidarians that is made of a web of interconnected neurons
The phylum that includes hydras, jellyfishes, corals, and sea anemones	The phylum that includes chitons, clams, snails, octopuses, squids, and mussels	A cell that beats a flagella to move water through a sponge	The mobile form of a cnidarian that is bell-shaped with the mouth and tentacles pointing down	Eyespots that detect light in cnidarians
The phylum that includes comb jellies	The phylum that includes sea stars, sea urchins, brittle stars, and sea cucumbers	A large hole at the top of a sponge	The jellylike material between the epidermis and gastrodermis in cnidarians	The sticky cells found on the two tentacles attached to ctenophores

5

10

15

8

66 *Biology Jeopardy*

©2004 Walch Publishing

The phylum that includes flatworms, flukes, and tapeworms	The phylum that includes ticks, spiders, scorpions, crabs, barnacles, centipedes, beetles, flies, and grasshoppers	The spiky particles made of calcium carbonate or silicon dioxide that give sponges structure	The internal cavity lined by the gastrodermis in which digestion occurs in cnidarians	The production of light by ctenophores
The phylum that includes rotifers	The subphylum of arthropods that lack antennae but have mouthparts called chelicerae	The cells that take up nutrients from the water and make spicules	Cells found in the tentacles of cnidarians that are used for defense and to capture prey	A class of cnidarians that contains Hydras, *Obelia*, and Portuguese man-of-wars
The phylum that includes pinworms, hookworms, and roundworms	The subphylum of arthropods that have two pairs of antennae and mouthparts called mandibles	Asexual reproductive structures made of amoebocytes (or archaeocytes) surrounded by spicules, allowing sponges to survive harsh conditions	Poison filaments released when cnidocytes are triggered by touch; they may have barbs, contain poison, or wrap around prey.	These structures are made from the calcium carbonate skeletons of polyps in the class Anthozoa that live in warm, clear waters.

20   **25**   **BONUS**   **NOTES**

Invertebrates
Section 27

	FLATWORMS I	FLATWORMS II	ROTIFERS AND ROUNDWORMS	ANNELIDS I	ANNELIDS II
5	A muscular tube used to bring food into the body of a flatworm	A class of flatworms that includes the free-living flatworms, such as planaria	The hairlike structures surrounding the mouth of a rotifer	The external bristle hairs attached to segments of the body	The structure used for food storage between the esophagus and gizzard
10	The cells that remove metabolic wastes, such as ammonia and urea, from a flatworm's body	A class of flatworms that includes parasitic flukes, such as the one that causes schistosomiasis in humans	The muscular organ beyond the mouth of rotifers that helps grind up food	The paddlelike appendages attached to body segments	The five pairs of muscular tubes that connect dorsal and ventral blood vessels
15	Clusters of nerve cells in the anterior region of a flatworm that control the nervous system	A class of flatworms that includes tapeworms	A parasitic roundworm that uses toothlike plates in the mouth to cut into skin and tissue, allowing it to feed on the blood of its host	The internal walls of tissue that separate body segments	A class of annelids that includes freshwater and terrestrial segmented worms such as earthworms

8

Biology Jeopardy

©2004 Walch Publishing

20	A group of cells used to detect changes in light	The head of an adult tapeworm that contains hooks or suckers and allows it to attach to a host	A disease that causes muscle pain and stiffness when undercooked meat containing roundworm larvae in cysts is eaten	Excretory structures used to eliminate waste products and excess fluid	A class of annelids that lack setae but have suckers at both ends of their bodies; includes leeches
25	A form of asexual reproduction used by flatworms to regenerate a missing piece of an organism	Segments behind the head of a tapeworm that contain both male and female reproductive structures	A disease caused by a filarial worm that blocks the passage of fluid in lymphatic vessels and causes swelling in surrounding tissue	The thickened band of segments that excretes mucus during reproduction	The muscular extension used by some leeches to enter the tissue of their host
BONUS	The body plan of a flatworm	Tapeworm larvae in a protective coating that can exist in muscle tissue	A roundworm that was the first multicellular organism to have its DNA sequenced completely	Earthworm feces that are rich in minerals and help maintain soil fertility	A class of annelids that have parapodia with setae; many of the species are marine and include sandworms and bloodworms.

NOTES

Invertebrates
Section 28

MOLLUSKS I	MOLLUSKS II	ARTHROPOD CHARACTERISTICS I	ARTHROPOD CHARACTERISTICS II	CRUSTACEANS I
The muscular organ used for locomotion	The hollow tubes that bring water into and out of filter feeders such as clams and scallops	Jointed structures, such as a leg or an antennae	A type of circulatory system found in arthropods	The free-swimming larval form of most crustaceans
A mass that contains internal organs, such as the heart, and organs for digestion, excretion, and reproduction	A class of mollusks that tend to have one shell or no shell; members of this class include snails, slugs, and nudibranchs.	A carbohydrate containing nitrogen found in the exoskeleton	A network of air tubes in terrestrial arthropods	Appendages in adult barnacles used to sweep food particles into the mouth
A layer of epidermal tissue that covers the visceral mass	A class of mollusks that have hinged shells divided into two halves; members of this class include clams, mussels, oysters, and scallops.	The process of shedding an old exoskeleton and replacing it with a new one	The opening to a tracheal tube	A group of small crustaceans found in marine and freshwater environments, making them an abundant member of plankton communities

5

10

15

8

A structure of calcium carbonate made by the mantle	A muscle used to hold together two halves of a bivalve to protect their soft bodies	The light-sensitive structures, each with its own lens	The folded respiratory structures in the abdomen of arachnids having a large surface area for gas exchange	A large group of marine species that also includes terrestrial species, such as pill bugs or wood lice
The tough, flexible band of tissue covered with teeth and used for feeding	A class of mollusks that tend to be soft-bodied with a foot divided into tentacles; members of this class include squid, octopuses, and nautiluses.	An extinct subphyla of arthropods that have many segments, each with a pair of appendages	The excretory organs in terrestrial arthropods that remove waste from blood and add it to feces	A group of crustaceans having five pairs of legs that include crayfish, lobsters, crabs, and shrimp
The larval stage of many aquatic mollusks	The only extant cephalopod with an external shell	An area in the Canadian Rockies where many multicellular fossils from the Cambrian Period have been found	One of the oldest species of arthropods having tough plates of armor and a spikelike tail	The region of a decapod consisting of the head and thorax

20

25

BONUS

NOTES

Biology Jeopardy 71

©2004 Walch Publishing

Invertebrates
Section 29

CRUSTACEANS II	CHELICERATES AND UNIRAMIANS	INSECTS I	INSECTS II	ECHINODERMS
5	The region of a decapod behind the cephalothorax	The first pair of appendages that act as pincers or contain fangs used to paralyze prey	The number of pairs of legs attached to the thorax	This occurs when an insect transfers pollen from one flower to another flower of the same species.	A network of water-filled canals used for respiration, circulation, and movement
10	An exoskeleton that covers the cephalothorax	The second pair of appendages used to hold and chew food	The mouthpart that acts like an upper lip	A type of developmental change that occurs when an egg hatches into a nymph and then develops into an adult	A water-filled circle that forms around the mouth
15	The appendages on the head of a decapod used as sense organs or feelers	The structures on the abdomen of arachnids used to spin silk into webs	The mouthpart that acts like a lower lip	A type of developmental change that occurs when an egg hatches into a larva, changes to a pupa, and then changes to an adult	A plate on the surface that connects to the water vascular system and allows water to enter echinoderms from outside the body

8

72 Biology Jeopardy ©2004 Walch Publishing

Extensions of the water vascular system with suckers on the end that aid in movement and pry open shells of bivalves	Calcium carbonate plates in the endoskeleton	A jawlike structure made of five teeth used by sea urchins for feeding	
A protective case formed around a developing butterfly pupa	A chemical used by insects to communicate with individuals of the same species	Insects, like bees, that live in a colony and divide work tasks among groups or castes	
A structure that secretes enzymes into the digestive tract	A structure in the female where sperm is deposited and stored until eggs are released by the ovaries	A structure in females used for laying eggs	
A class of chelicerates that includes spiders, mites, ticks, and scorpions	A type of uniramian that has two pairs of legs attached to most body segments	A type of carnivorous uniramian that has one pair of legs attached to most body segments	
A pair of appendages used for biting and grinding food	The first pair of appendages on a decapod that are enlarged and used to capture, crush, or cut food	The appendages attached to the abdomen of a decapod used for swimming and reproduction	
20	**25**	**BONUS**	**NOTES**

Biology Jeopardy 73

UNIT 9

Vertebrates

Vertebrates
Section 30

	Chordates and Nonvertebrate Chordates	Vertebrate Classification	Fishes I	Fishes II	Fishes III
5	A hollow tube on the back of vertebrates that develops into the brain and spinal cord	A class containing humans, dogs, cats, and horses; these organisms have hair and produce milk in mammary glands for their young.	A class of jawless fishes	The overlapping plates covering the skin of most fish that serve as protection	A sac containing gases in bony fish that allows them to be buoyant
10	A pair of pockets off the pharynx in chordates	A class containing owls, hawks, jays, and sparrows; these organisms have adaptations for flying such as feathers and hollow bones.	A type of jawless fish with short tentacles around the mouth that lives in marine waters and tends to burrow into and eat dead fish	A feathery structure located on both sides of the pharynx used for gas exchange in most fish	A type of development in which eggs are laid and young develop and hatch outside the body of the female
15	A flexible supporting rod found below the nerve cord in chordates	A class containing lizards, snakes, turtles, alligators, and crocodiles; these organisms have scaly skin and tend to lay their eggs on land.	A type of jawless fish that lives in freshwater and has a disklike mouth with teeth used to attach to and scrape a hole in a host	A plate that covers and protects the gills in bony fish	A type of development in which eggs remain in the female, young derive nutrients from the egg yolk, and then are born alive

20	A type of tail found in chordates that is made of bone and muscle	A class containing salamanders, frogs, and toads; these organisms have thin skin, live in moist places, and tend to lay their eggs in water.	The first portion of the heart that receives oxygen-poor blood from the body	A fin that extends from the tail of a bony fish	A type of development in which young develop in and derive nutrients from the placenta in the female and are born alive
25	A blade-shaped nonvertebrate chordate that lives on sandy bottoms of warm shallow waters	A class containing trout, salmon, perch, and bass; these organisms have jaws and a bony endoskeleton.	A form of circulation that takes blood from the heart to the gills to the rest of the body and back to the heart	The bone structure located behind the operculum of bony fish to which the pectoral fin is attached	Fish that live in marine environments but migrate to freshwater to breed
BONUS	A barrel-shaped nonvertebrate chordate that is covered with a tough outer covering and tends to be sessile as an adult	A class containing sharks, skates, and rays; these organisms have jaws and an endoskeleton made of cartilage.	The process in the body of a fish that moves blood toward the head while water moves away from the head toward the gills	A row of sensory receptors found on both sides of most fish to detect motion and currents	Fish that live in freshwater but migrate to marine environments to breed

NOTES

Vertebrates
Section 31

	AMPHIBIANS I	AMPHIBIANS II	REPTILES I	REPTILES II	BIRDS I
5	Amphibian larvae that swim and have a tail	The muscular sac at the end of the large intestine through which waste, urine, egg, and sperm pass before leaving the body	The sac that contains fat and nutrients for a developing embryo in an egg	The protein that fills cells to produce tough, scaly skin	The tough covering on the jaw of a bird that is adapted for many feeding styles
10	The glands that help keep the skin of amphibians moist and lubricated	The transparent membrane found under the eyelid that moves to protect and moisten the eye	The fluid-filled sac that protects and cushions the embryo	Extinct reptiles that lived during the Mesozoic Era	The modified forelimb used for flight in most birds
15	The gas exchange that occurs across the skin	The external eardrums located on both sides of the head	The structure that stores nitrogenous waste in an egg and allows for gas exchange when it fuses with the chorion	The order that includes crocodiles, alligators, and caimans	A fluffy feather that traps air close to the body for insulation

20	The number of chambers in an amphibian heart	The order of amphibians that contains frogs and toads	The sense organ at the top of the mouth that detects chemicals and perceives odors	The order that includes turtles and tortoises	The outer feather that streamlines the shape of the bird and adds color and insulation
25	The process by which a male climbs onto the back of a female and holds her in an embrace until she lays eggs	The order of amphibians that contains salamanders	The top or dorsal portion of a turtle or tortoise shell	The order that includes lizards and snakes	The skeletal adaptation for flight
BONUS	The vertebrae at the end of the spine that are fused together	The order of amphibians that contains legless caecilians	The bottom or ventral portion of a turtle or tortoise shell	The order consisting of only two living species of tuataras	The oldest known bird fossil that shows characteristics of both birds and reptiles

NOTES

Vertebrates
Section 32

	BIRDS II	BIRD ORDERS	MAMMALS I	MAMMALS II	MAMMAL ORDERS
5	The muscular organ beyond the crop that helps grind food	The largest order of birds, which includes robins, warblers, and jays—also known as songbirds	The glands that produce milk to nourish young	The number of chambers in a mammal heart	The order of mammals with five-digit hands, good binocular vision, and well-developed brains; includes humans, monkeys, and chimpanzees
10	The large breastbone to which flight muscles are attached	Also known as raptors, members of this group, such as hawks, eagles, vultures, and falcons, have sharp, curved beaks and sharp talons.	An organism that generates and regulates its body temperature internally	The process of bouncing sound waves off an object to determine the distance to an object	The order of mammals that are carnivores and includes bears, seals, dogs, and raccoons
15	The location where most air goes during respiration before going to the lungs	The group of birds that includes woodpeckers and toucans; all members of the group nest in tree cavities.	The flat muscle below the rib cage that contracts and relaxes and, in the process, controls the lung volume	A mammal that lays eggs and then feeds mother's milk to the young	The largest order of mammals; mostly herbivores but may be omnivores with sharp incisors, such as squirrels, mice, rats, beavers, and porcupines

	The order of mammals adapted to live and give birth in the water, such as whales, dolphins, and porpoises	The order of mammals that fly, are normally nocturnal, and use echolocation, such as bats	The order of mammals that are hoofed herbivores with two digits on each foot, such as cows, sheep, antelopes, and pigs	
	A mammal that gives birth to young shortly after conception; the young completes development in an external pouch.	A mammal that gives birth to live young after an extended developmental period in the female, during which a placenta is used to nourish the embryo	Mammals with hooves	
	The first chamber of the stomach that uses microorganisms to break down plant matter in animals such as cows	The front, chisellike teeth used for cutting and gnawing food	The back teeth used for crushing and grinding food	
	This group includes large-bodied birds known as fowl, such as chickens, turkeys, and grouse.	This group includes owls, mostly nocturnal predators that have sharp beaks and talons.	This group includes the waterfowl that have webbed feet and flattened bills, such as ducks, geese, and swans.	
	The process by which birds use their beaks to spread oil on their feathers	The movement of species from one location to another based on a change in seasons	The thickened, featherless patch of skin on the abdomen of birds used to incubate eggs while sitting on them	
20		**25**	**B O N U S**	**N O T E S**

©2004 Walch Publishing

UNIT 10

Human Biology

Human Biology
Section 33

	Tissues and Organ Systems I	Tissues and Organ Systems II	The Skeleton I	The Skeleton II	Bone Structure
5	This system supports the body and protects internal organs; it serves as a site for mineral storage and red blood cell formation.	This system breaks down food into smaller molecules and absorbs nutrients.	The part of the skeleton consisting of the skull, vertebral column, sternum, and rib cage	The bone structure protecting the brain	A latticework structure of bone found at the end of long bones and in the middle of shorter, flat bones
10	This system functions to support and move the body and helps to move both blood and food through the body.	This system removes wastes from the body using structures like the kidney, skin, and lungs.	The part of the skeleton consisting of the bones in the arms, legs, pelvis, and shoulder area	A bone located in the upper arm	A dense bone found on the outside of bones
15	The tightly packed layers of cells that line all body surfaces and organs	This system helps the body respond to changes in its internal and external environment using the brain, spinal cord, nerves, and sense organs.	Located in the thigh, this is the longest bone in the body.	A flat bone located at the center of the chest	The location in the bone where red blood cells are made

84 Biology Jeopardy   ©2004 Walch Publishing

The type of bone marrow made mostly of fat cells	Blood vessels and nerves run through these narrow tubes in bones.	A tough membrane of connective tissue on the outside of the bone that contains blood vessels and nerves	
One of two bones found in the lower leg	Another name for the collarbone	The location of the carpals	
The bones found in the fingers and toes	The location of the patella	The number of bones found in the adult human body	
The tissue that binds body structures together and holds organs in place	This system produces sperm in males and eggs and milk in females and can nourish an embryo after fertilization.	This system controls homeostasis, metabolism, growth and sexual development, and maintains water and mineral balance.	
The major structures of this system include skin, hair, and nails.	This system carries air into and out of the lungs where oxygen and carbon dioxide are exchanged.	The major structures of this system include the heart, blood vessels, and blood.	
20	25	BONUS	NOTES

Biology Jeopardy 85

Human Biology
Section 34

Bone Formation	Joints	The Muscular System	Muscle Movement	The Integumentary System
A newborn baby's skeleton is made mostly of this tough, flexible connective tissue.	The type of joint found between the femur and pelvis that allows for circular movement of the leg	Striated muscle containing either one or two nuclei per cell found only in the heart	A thick band of connective tissue that attaches skeletal muscle to bone	The outer layer of skin, made of both dead and living cells
The process of replacing cartilage with minerals to create bone	The type of joint found between plates in the skull	Striated, voluntary muscle made of long cells containing many nuclei	The point at which a skeletal muscle attaches to a stationary bone	The inner layer of skin containing blood vessels, nerve cells, muscle fibers, hair follicles, and glands
Cells that help monitor and maintain cellular activity in bones	A band of tough connective tissue attaching one bone to another bone	Non-striated, involuntary muscle made of spindle-shaped cells containing one nucleus per cell	The point at which a skeletal muscle attaches to a moving bone	A pocket of cells where hair is produced

10

5

10

15

Cells that make new bone for growth and repair	A lubricating substance at the end of bones that reduces friction between bones	Thick filaments of protein found in skeletal muscle	A muscle that bends a joint, such as a biceps muscle	A protein found in the upper layer of skin cells, hair cells, and nail cells
Cells that break down and dissolve bone	Small sacs filled with synovial fluid found in joints	Thin filaments of protein found in skeletal muscle	A muscle that straightens a joint, such as a triceps muscle	A brown pigment found in skin cells
The site of bone elongation at the end of long bones like the femur	A disorder that causes swelling and pain in joints	The region between Z discs in a skeletal muscle fiber	A contact point between a neuron and a muscle cell	A gland that produces an oily secretion or sebum

20

25

BONUS

NOTES

©2004 Walch Publishing

Biology Jeopardy 87

Human Biology
Section 35

10

	THE DIGESTIVE SYSTEM I	THE DIGESTIVE SYSTEM II	THE DIGESTIVE SYSTEM III	THE EXCRETORY SYSTEM I	THE EXCRETORY SYSTEM II
5	The first place in the body where mechanical digestion occurs	A muscular sac in the body where mechanical and chemical digestion occur	A large organ located below the diaphragm that stores glycogen and breaks down toxic substances, such as alcohol	The main organ of the excretory system found in the lower back	The small functional unit of a kidney
10	Glands that aid with chemical digestion in the mouth	A ground-up mixture of food and gastric fluid found in the stomach	Made in the liver, this fluid helps to break down fats or lipids.	A muscular sac that stores urine	A small bundle of capillaries enclosed by a cup-shaped structure in a nephron
15	A muscular tube that connects the pharynx to the stomach	The enzyme that breaks down large proteins in the stomach	A pouchlike organ where bile is stored	A tube that takes urine from the kidney to the urinary bladder	A cup-shaped structure surrounding capillaries in a nephron

The section of the nephron where sodium chloride is concentrated in the surrounding fluid around the nephron and water is conserved in the body	The process of removing water, urea, glucose, vitamins, and salts at the Bowman's capsule	The process of transporting or returning needed materials from the filtrate to the blood	
A tube that takes urine out of the body from the urinary bladder	The name of the outer portion of the kidney	The toxic nitrogenous waste made in the liver that is taken in the bloodstream to the kidney where it is eliminated from the body	
The organ that releases enzymes for digestion, sodium bicarbonate, and hormones to regulate blood sugar levels	The final segment of the digestive system in which water can still be removed from the undigested material	Organisms found in the large intestine that help the body make vitamins, such as vitamin K	
A ring of muscle found between the stomach and small intestine	The first section of the small intestine	Small projections lining the small intestine that help with the absorption of nutrients	
The name given to a moistened ball of food that enters the pharynx	Smooth muscle contractions that push food through the esophagus	A thick ring of muscle that opens and closes between the esophagus and the stomach	
20	**25**	**BONUS**	**NOTES**

Human Biology
Section 36

	Food and Nutrition	The Human Heart I	The Human Heart II	Blood Vessels	Blood
5	The amount of heat needed to raise the temperature of one gram of water by one Celsius degree	The number of chambers found in a human heart	The largest artery in the body	The smallest blood vessels in the human body	The straw-colored fluid that makes up the majority of the blood
10	Sugars and starches are examples of these organic compounds that contain carbon, hydrogen, and oxygen in a one-to-two-to-one ratio.	The top chambers of the heart that receive blood	A group of cells in the right atrium that regulate the heartbeat	Thick-walled blood vessels that almost always carry oxygen-rich blood	The most common type of cell in the blood
15	An organic compound used to store energy reserves and insulate the body; saturated, unsaturated, and polyunsaturated are all examples of these.	The bottom chambers of the heart that pump blood out of the heart	The pathway that takes oxygen-poor blood from the right side of the heart to the lungs	Blood vessels containing valves that almost always carry oxygen-poor blood	The blood cells that attack foreign entities in the body, such as bacteria

Organic compounds made of amino acids	A flap of tissue in the heart that prevents blood from flowing backward	The pathway that takes oxygen-rich blood from the heart to all parts of the body except the lungs	The measurement of the force of blood against the artery wall	An iron-containing protein that binds oxygen to red blood cells
Organic molecules that can be water-soluble or fat-soluble and are used to help enzymes function in the body	A thick wall of muscle that separates the right side of the heart from the left side of the heart	The blood vessel that brings oxygen-poor blood from the head and arms to the right atrium	The first measurement in blood pressure when the ventricle contracts	Cell fragments that aid in blood clotting
Inorganic nutrients, such as iron, calcium, and magnesium, that help the body function	A protective membrane that surrounds the heart	Blood vessels that bring oxygen-rich blood from the lungs back to the left atrium	The second measurement in blood pressure when the ventricle relaxes	A protein found in plasma that helps clot blood

20

25

BONUS

NOTES

©2004 Walch Publishing

Biology Jeopardy 91

Human Biology
Section 37

THE CIRCULATORY SYSTEM AND BLOOD DISORDERS	BLOOD TYPES	THE LYMPHATIC SYSTEM	THE RESPIRATORY SYSTEM I	THE RESPIRATORY SYSTEM II
5 — A condition that results if an artery in the heart is blocked and heart tissue is deprived of oxygen	Proteins found on the surface of red blood cells that determine blood type	The excess fluid that moves into tissue from the blood	A tube at the back of the mouth and nasal passage through which both air and food pass	During respiration, this gas passes from the air into the lungs through the capillaries surrounding the alveoli and to the blood.
10 — A condition that results in fatty deposits or plaque buildup on the walls of arteries	The blood type of a person who has both A antigens and B antigens on their red blood cells	A small clump of lymph vessels where lymph is filtered	Also known as the windpipe, it is a tube made of cartilage.	During respiration, this gas passes from the blood through the capillaries surrounding the alveoli and into the lungs.
15 — A condition that results when blood pressure is too high	The blood type of a person with no antigens on their red blood cells	White blood cells found in the lymph nodes that help the body fight infection	A piece of cartilage that covers the opening to the trachea	A large, flat skeletal muscle that separates the chest cavity from the abdominal cavity and works by causing a change of air pressure in the lungs

Biology Jeopardy

20	A condition that results when a blood clot, formed as a result of atherosclerosis, breaks away from an artery and lodges in the brain	Proteins made in response to foreign antigens in the body	Located behind the stomach, this organ made of lymphatic tissue helps break down damaged red blood cells and platelets	Located at the top of the trachea, it contains the vocal cords.	A poisonous gas found in tobacco smoke that binds with hemoglobin and blocks oxygen from binding with the hemoglobin in red blood cells
25	A genetic disorder that results in a faulty blood-clotting protein and reduces the rate of blood clotting in the body	The name given to the clumping of blood when the different blood types are mixed together	A swelling in the tissue that can occur when fluid builds up due to a blockage in lymphatic vessels	Large tubes made of cartilage and smooth muscle that allow air to pass from the trachea to the lungs	A respiratory disease caused by the inflammation of bronchi or bronchioles
BONUS	A condition that is characterized by a very high level of white blood cells circulating in the body	Named for the rhesus monkey in which it was discovered, this antigen is sometimes present on the surface of red blood cells.	The location in the circulatory system where lymph is returned to the bloodstream	Air sacs at the end of the bronchioles in which gas exchange occurs	A respiratory disease that causes the alveoli and other lung tissues to lose their elasticity

NOTES

Human Biology
Section 38

	The Nervous System I	The Nervous System II	The Neuron	The Brain	The Senses I
5	A division of the nervous system that includes the brain and spinal cord	An unconscious movement that normally serves to protect the body from harm	The largest part of the neuron that contains a nucleus and most of the cytoplasm	The largest part of the brain that is responsible for conscious activities	A sensory receptor that responds to changes in temperature
10	A division of the nervous system that includes all of the neurons not found in the brain and spinal cord	The location between two neurons where information is transmitted from one neuron to the other	Long extensions of the cell body that receive electrical signals from other neurons	Located at the back of the skull, this part of the brain helps with balance and coordination.	A sensory receptor found in the nose and taste buds
15	A division of the peripheral nervous system controlling voluntary movement of skeletal muscles and involuntary actions such as reflexes	Chemicals released at the axon terminal of one neuron into a synapse with another neuron	A long extension that carries electrical impulses away from the cell body	The region found under the cerebellum that helps to relay information between the brain and the rest of the body	The opening in the eye that adjusts in size to let in more or less light

94 Biology Jeopardy

©2004 Walch Publishing

The membrane in the ear that vibrates in responses to changes in air pressure	A receptor found on the retina that detects color	The tubes found in the inner ear that help the body maintain balance	
A band of neural axons that connects the left and right hemisphere of the brain	Three layers of membranes that protect the brain and spinal cord	The back section of the cerebrum that helps to process visual information	
A lipid layer that insulates the axon to help speed the transmission of an electrical impulse the length of the neuron	A neuron that transmits information from the external or internal environment to the brain or spinal cord	A neuron that transmits information from the brain or spinal cord to muscles or glands	
The difference in charge across the membrane of a neuron when the charge is negative inside and positive outside the neuron	A change in charge across the membrane of a neuron when the charge is positive inside and negative outside the neuron	The time between the beginning of an action potential and the next resting potential; another impulse cannot be transmitted during this time.	
A division of the peripheral nervous system that controls involuntary movement of smooth muscles regulating internal mechanisms	A division of the autonomic nervous system that directs blood to the heart and skeletal muscles under stress; also known as "fight-or-flight" response	A division of the autonomic nervous system that helps the body conserve energy and returns the body to normal conditions after a period of stress	
20	25	BONUS	NOTES

Biology Jeopardy

Human Biology
Section 39

10

THE SENSES II	DRUGS AND THE NERVOUS SYSTEM	THE ENDOCRINE SYSTEM	ENDOCRINE GLANDS	HORMONES I
A sensory receptor that responds to touch, pressure, and movement	When a user becomes dependent or cannot function comfortably without a drug	A chemical compound that travels from its source to the rest of the body	Located below the brain, this gland is controlled by the hypothalamus.	In males, the sex hormone that stimulates the development of secondary sex characteristics and the production of sperm
A tough, clear layer on the outer surface of the eye	A drug that increases the action of the central nervous system	A gland that secretes nonhormonal compounds from the inside to the outside of the body	Located below the larynx, this gland helps control the body's metabolism.	A hormone containing iodine that helps to regulate the body's metabolism
A fluid-filled structure in the inner ear that is lined with hairs that respond to changes in pressure	A drug that decreases the action of the central nervous system	A gland that secretes hormones into the bloodstream	Located behind the stomach, this gland helps regulate sugar levels in the blood.	A hormone that stimulates the cells in the body to absorb glucose from the bloodstream

5

10

15

96 Biology Jeopardy

©2004 Walch Publishing

One of the two hormones made in the adrenal medulla that helps the body cope with stress	A hormone that stimulates the kidneys to reabsorb water	A hormone that stimulates uterine contraction during childbirth and the flow of breast milk in nursing mothers	
Located in the cerebrum, this gland helps regulate sleep cycles.	Located above the kidneys, these glands help control the body's reaction to stress.	Located under the sternum, these glands play a role in the development of the immune system.	
A hormone made of lipids that can pass through cell membranes	A hormone made of amino acids that cannot pass through cell membranes	Hormones released by cells that act on surrounding or nearby cells	
A group of birth defects caused by alcohol use in pregnant women	An illegal stimulant that causes the brain to release dopamine	A group of drugs that mimic natural painkillers in the body	
A receptor in the nose that detects chemical cues	The structure that relays visual images from the eye to the brain	A jellylike substance found behind the lens in the eye	
20	**25**	**BONUS**	**NOTES**

Human Biology
Section 40

	HORMONES II	ENDOCRINE SYSTEM DISORDERS	THE MALE REPRODUCTIVE SYSTEM	THE FEMALE REPRODUCTIVE SYSTEM	THE MENSTRUAL CYCLE
5	In females, one of the two sex hormones that stimulates the development of secondary sex characteristics and production of eggs	A disorder in which too little insulin is produced; there is a high level of glucose in the blood.	Male gametes or sex cells	The female gonad that produces sex cells or eggs	The period when a follicle matures to nourish an egg and the estrogen level increases
10	A hormone that stimulates the liver to convert glycogen into glucose so it can be released into the bloodstream	A disorder in which too much insulin is produced; there is too little glucose in the blood.	The external organ that leads the urethra to the outside of the body	The muscular canal that leads from the cervix to the outside of the body	The shortest period that occurs when a follicle ruptures and an egg is released
15	A hormone that regulates the sleep-wake cycle	A condition that results when too many thyroid hormones are produced	The location where sperm are produced	The muscular organ in which a fertilized egg may implant and develop	The period when the follicle changes to the corpus luteum; it secretes large amounts of progesterone and estrogen.

10

98   Biology Jeopardy   ©2004 Walch Publishing

A hormone that increases the amount of calcium in the blood by causing the kidneys to reabsorb calcium into the bloodstream	A condition that results when too few thyroid hormones are produced	The combination of sperm and seminal fluid	A narrow tube that starts near the ovary and extends to the uterus	The period when the uterine lining is shed if fertilization does not occur
A hormone that decreases the amount of calcium in the blood by causing the kidneys to not reabsorb as much calcium into the bloodstream	A disorder caused by an iodine deficiency that can result in an enlarged thyroid	The structure in which sperm mature and are stored	The lower end of the uterus that is the opening to the vagina	The stage that occurs when all of the follicles have matured or disintegrated and menstruation ceases
A hormone that stimulates the production of milk in nursing mothers	A disorder that can develop in childhood if iodine is lacking in the diet; as a result, the skeletal and nervous systems do not develop correctly.	A tube that takes mature sperm from the epididymis to the urethra	The fingerlike projections at the end of the Fallopian tubes that help sweep a mature egg into the Fallopian tube	The name of one of the two hormones made in the pituitary gland that are involved in the menstrual cycle

20

25

BONUS

NOTES

Biology Jeopardy

Human Biology
Section 41

	FERTILIZATION AND DEVELOPMENT I	FERTILIZATION AND DEVELOPMENT II	THE IMMUNE SYSTEM I	THE IMMUNE SYSTEM II	CANCER
5	The process of a sperm fusing with an egg	The name given to an embryo after the eighth week of fertilization until birth	An organism that causes a disease	A reaction to harmless antigens such as pollen and dander	A mass of tissue that results when cells divide abnormally
10	The name of a fertilized egg	It connects the fetus to the placenta and contains both arteries and veins.	A rise in body temperature above 37 degrees C in response to an infection	The chemical released by injured cells to dilate surrounding blood vessels	A tumor that is noncancerous and does not spread to other tissue in the body
15	The name given to an embryo that is a solid ball of cells	The membrane that forms a fluid-filled covering surrounding the embryo or fetus	When a pathogen breaks through the skin or a mucous membrane and white blood cells come to that area	Cells infected by a virus make these proteins that protect other cells from viral infection.	When a tumor grows quickly and spreads to healthy tissue

The name given to an embryo that is a hollowed-out ball filled with fluid	The structure in the uterus that allows an exchange of gases, nutrients, and waste products between a mother and a fetus	The reaction of the body against a pathogen	A resistance to specific pathogens
The process of the blastocyst attaching to the lining of the uterus	The membrane that becomes the placenta	The type of reaction involving T-cells attacking cells bearing antigens	A weakened form of a pathogen that triggers an immune system response
The process that occurs when an embryo develops three germ layers—the ectoderm, mesoderm, and endoderm	The muscular contractions and hormonal changes that lead to childbirth	The type of reaction to pathogens in the body fluids using antibodies and B cells	A disorder in which the immune system attacks its own body cells
			The spread of cancer cells from one site in the body to another
			A chemical substance that causes cancer
			The procedure of using chemicals to destroy cancer cells

20

25

BONUS

NOTES

©2004 Walch Publishing

Biology Jeopardy 101

Answer Key

1 Introduction to Biology
Section 1

	The Process of Science	Characteristics of Life I	Characteristics of Life II	Common Metric Units	Metric Unit Conversions
5	What is an observation?	What is a cell?	What is a stimulus?	What is a gram?	What is 1,000?
10	What are data?	What is a unicellular organism?	What is a response?	What is a meter?	What is 1,000?
15	What is an inference?	What is a multicellular organism?	What is evolution?	What is a liter?	What is 100?
20	What is a hypothesis?	What is sexual reproduction?	What is metabolism?	What is a second?	What is 10?
25	What is a theory?	What is asexual reproduction?	What is homeostasis?	What is a kelvin?	What is 1,000,000?
BONUS	What is a controlled experiment?	What is development?	What is a gene?	What is an ampere?	What is 1,000?

1 Introduction to Biology
Section 2

	Microscopes	Branches of Biology I	Branches of Biology II	Branches of Biology III	The Composition of Matter I
5	What is a compound light microscope?	What is zoology?	What is botany?	What is virology?	What is an electron?
10	What is the stage?	What is mammalogy?	What is ecology?	What is mycology?	What is a proton?
15	What are objective lenses?	What is paleontology?	What is microbiology?	What is herpetology?	What is a neutron?
20	What is an electron microscope?	What is ornithology?	What is ethology?	What is lepidoptery?	What is atomic number?
25	What is a transmission electron microscope?	What is ichthyology?	What is genetics?	What is embryology?	What is mass number (or atomic mass)?
BONUS	What is a scanning electron microscope?	What is cytology?	What is entomology?	What is histology?	What are isotopes?

1 Introduction to Biology
Section 3

	The Composition of Matter II	Properties of Water I	Properties of Water II	Elements in Organic Compounds	Organic Compounds I
5	What is an element?	What is a mixture?	What is cohesion?	What is carbon?	What are monomers?
10	What is a compound?	What is a solution?	What is adhesion?	What is hydrogen?	What is a polymer?
15	What is a molecule?	What is a solute?	What is capillary action?	What is oxygen?	What is condensation or dehydration synthesis?
20	What is an ion?	What is a solvent?	What is an acid?	What is nitrogen?	What is hydrolysis?
25	What is an ionic bond?	What is a polar molecule?	What is a base?	What is phosphorous?	What are isomers?
BONUS	What is a covalent bond?	What is a hydrogen bond?	What is a buffer?	What is sulfur?	What is a covalent bond?

1 Introduction to Biology
Section 4

	Organic Compounds II	Organic Compounds III	Organic Compounds IV	Organic Compounds V	Organic Compounds VI
5	What is a carbohydrate?	What is maltose?	What is a lipid?	What are nucleic acids?	What is a catalyst?
10	What is a monosaccharide?	What is a polysaccharide?	What is a saturated fatty acid?	What is deoxyribonucleic acid (DNA)?	What is an enzyme?
15	What is fructose or galactose?	What is starch?	What is an unsaturated fatty acid?	What is ribonucleic acid (RNA)?	What is activation energy?
20	What is a disaccharide?	What is glycogen?	What is a polyunsaturated fatty acid?	What is a protein?	What are reactants?
25	What is sucrose?	What is cellulose?	What is glycerol?	What is twenty?	What are products?
BONUS	What is lactose?	What is chitin?	What is a phospholipid?	What is a peptide bond?	What is a substrate?

Ecology — Section 5

	Levels of Organization	Obtaining Energy in an Ecosystem	The Flow of Energy in an Ecosystem	Ecosystem Relationships	Chemical Cycles
5	What is an organism?	What is an autotroph (or producer)?	What is a food chain?	What is habitat?	What is the water cycle?
10	What is a population?	What is a heterotroph (or consumer)?	What is a food web?	What is a niche?	What is the carbon cycle?
15	What is a community?	What is an herbivore?	What is a trophic level?	What is predation?	What is the nitrogen cycle?
20	What is an ecosystem?	What is a carnivore?	What is an energy pyramid?	What is mutualism?	What is the phosphorous cycle?
25	What is a biome?	What is an omnivore?	What is a pyramid of numbers?	What is parasitism?	What is a biogeochemical cycle?
BONUS	What is the biosphere?	What is a decomposer?	What is a biomass pyramid?	What is commensalism?	What is a limiting nutrient?

Ecology — Section 6

	Terrestrial Biomes	Aquatic Ecosystems	Population Change	Ecosystem Change I	Ecosystem Change II
5	What is a tropical rainforest?	What is an estuary?	What is immigration?	What is primary succession?	What is pollution?
10	What is a desert?	What is a wetland?	What is emigration?	What is secondary succession?	What is acid rain?
15	What is a temperate deciduous forest?	What is the photic zone?	What is exponential growth?	What is a pioneer species?	What is an endangered species?
20	What is the taiga or boreal forest?	What is the intertidal zone?	What is logistic growth?	What is desertification?	What is extinction?
25	What is the tundra?	What are coral reefs?	What is a density-dependent limiting factor?	What is deforestation?	What is global warming?
BONUS	What is a temperate grassland?	What is the benthic zone?	What is a density-independent limiting factor?	What is the ozone layer?	What is biodiversity?

Cell Biology
Section 7

	Cell Discoveries	Cell Structures I	Cell Structures II	Cell Structures III	Cellular Organization
5	Who is Robert Hooke?	What is the nucleus?	What is a chloroplast?	What is the nuclear envelope?	What are organelles?
10	Who is Anton van Leeuwenhoek?	What is the cytoplasm?	What are mitochondria?	What is a chromosome?	What is tissue?
15	Who is Matthias Schleiden?	What is the cell membrane?	What is a vacuole?	What is the nucleolus?	What is an organ?
20	Who is Theodor Schwann?	What is the cell wall?	What are lysosomes?	What are microtubules?	What is an organ system?
25	Who is Rudolf Virchow?	What are ribosomes?	What is the rough endoplasmic reticulum?	What are microfilaments?	What is a prokaryotic cell?
BONUS	Who is Lynn Margulis?	What is the Golgi apparatus?	What is the cytoskeleton?	What is the smooth endoplasmic reticulum?	What is a eukaryotic cell?

Cell Biology
Section 8

	The Cell Membrane I	The Cell Membrane II	Photosynthesis and Energy Carriers	Photosynthesis I	Photosynthesis II
5	What is diffusion?	What is facilitated diffusion?	What is adenosine triphosphate or ATP?	What is sunlight?	What is the thylakoid membrane?
10	What is selective permeability?	What is active transport?	What is two?	What is carbon dioxide?	What is a granum?
15	What is osmosis?	What is endocytosis?	What is ribose?	What is oxygen?	What is photosystem II?
20	What is isotonic?	What is exocytosis?	What is ATP synthase?	What are carbohydrates (or sugars and starches)?	What is photosystem II?
25	What is hypotonic?	What is turgor pressure?	What is chemiosmosis?	What is chlorophyll a?	What is the electron transport chain?
BONUS	What is hypertonic?	What is plasmolysis?	What is NADPH?	What is chlorophyll b?	What is photosystem I?

Cell Biology
Section 9

	Photosynthesis III	Cellular Respiration I	Cellular Respiration II	Cellular Respiration III	Cell Reproduction I
5	What is the Calvin Cycle (or light-independent reaction)?	What is aerobic?	What is 2?	What is Coenzyme A?	What is DNA?
10	What is the stroma?	What is anaerobic?	What is pyruvic acid?	What is acetyl-CoA?	What are histones?
15	What is RuBP?	What are the mitochondria?	What is NADH?	What is oxaloacetic acid?	What is a chromosome?
20	What is PGA?	What is glycolysis?	What is alcoholic fermentation?	What is citric acid?	What is a sister chromatid?
25	What is PGAL?	What is glucose?	What is lactic acid fermentation?	What is the electron transport chain?	What is a centromere?
BONUS	What is the C_4 pathway?	What is the cytoplasm?	What is the Krebs cycle?	What is ATP synthase?	What is the cell cycle?

Cell Biology
Section 10

	Cell Reproduction II	Cell Reproduction III	Cell Reproduction IV	Cell Reproduction V	Cell Reproduction VI
5	What is interphase?	What is mitosis (or the M phase)?	What is a cleavage furrow?	What is meiosis?	What is crossing-over?
10	What is the S phase (or synthesis phase)?	What is metaphase?	What is a cell plate?	What are homologous chromosomes?	What is prophase I?
15	What is the gap 2 phase (or G_2 phase)?	What is prophase?	What are daughter cells?	What is diploid?	What is anaphase I?
20	What is the gap 1 phase (or G_1 phase)?	What is anaphase?	What are cyclins?	What is haploid?	What is metaphase II?
25	What are centrioles?	What is telophase?	What is cancer?	What is spermatogenesis?	What is anaphase II?
BONUS	What are spindle fibers?	What is cytokinesis?	What is a tumor?	What is oogenesis?	What is telophase II?

Genetics
Section 11

	Mendelian Genetics I	Mendelian Genetics II	Mendelian Genetics III	Genetic Crosses I	Genetic Crosses II
5	Who is Gregor Mendel?	What is self-pollination?	What is dominant?	What is probability?	What is a monohybrid cross?
10	What are pea plants?	What is cross-pollination?	What is recessive?	What is a Punnett square?	What is a dihybrid cross?
15	What is heredity?	What are true breeding?	What is short?	What is homozygous?	What is a testcross?
20	What is genetics?	What is the first filial generation (or F_1 generation)?	What is green?	What is heterozygous?	What is 100% tall?
25	What are traits?	What is a gamete?	What is wrinkled?	What is phenotype?	What is 100% heterozygous?
BONUS	What is an allele?	What is the law of segregation?	What is the law of independent assortment?	What is genotype?	What is 100% round and yellow?

Genetics
Section 12

	Nucleic Acids I	Nucleic Acids II	Nucleic Acids III	Nucleic Acids IV	Nucleic Acids V
5	What is deoxyribonucleic acid (DNA)?	What is deoxyribose?	What is a double helix?	What is ribose?	What is RNA polymerase?
10	What is transformation?	What is a purine?	What are hydrogen bonds?	What is uracil?	What is translation?
15	Who is Oswald Avery?	What is a pyrimidine?	What is replication?	What is messenger RNA (mRNA)?	What is a ribosome?
20	Who are Martha Chase and Alfred Hershey?	What is adenine?	What is a replication fork?	What is ribosomal RNA (rRNA)?	What is a codon?
25	Who is Rosalind Franklin?	What is cytosine?	What is helicase?	What is transfer RNA (tRNA)?	What is an anticodon?
BONUS	Who are James Watson and Francis Crick?	What is thymine?	What is DNA polymerase?	What is transcription?	What are amino acids?

Genetics
Section 13

	Gene Expression	Genetic Errors	Inheritance Patterns	Human Inheritance I	Human Inheritance II
5	What is a promoter?	What is a mutation?	What are sex-linked traits?	What is a pedigree?	What is the Y chromosome?
10	What is an intron?	What is a point mutation?	What are multiple alleles?	What is a carrier?	What is monosomy?
15	What is an exon?	What is a frameshift mutation?	What is codominance?	What are sex chromosomes?	What is trisomy?
20	What is an operon?	What is a deletion mutation?	What is incomplete dominance?	What are autosomes?	What is Down syndrome?
25	What is a repressor protein?	What is an inversion mutation?	What are polygenic traits?	What is a karyotype?	What is Turner's syndrome?
BONUS	What is an operator?	What is a translocation mutation?	What is nondisjunction?	What is a Barr body?	What is Klinefelter's syndrome?

Genetics
Section 14

	Human Inheritance III	Human Inheritance IV	DNA Research I	DNA Research II	DNA Research III
5	What is colorblindness?	What is albinism?	What is a fruit fly (or *Drosophilia melanogaster*)?	What is genetic engineering?	What is a primer?
10	What is hemophilia?	What is sickle-cell anemia/disease?	Who is Thomas Hunt Morgan?	What is recombinant DNA?	What is cloning?
15	What is Duchenne Muscular Dystrophy?	What is Huntington's disease?	What is a gene map?	What is a plasmid?	What is Dolly?
20	What is cystic fibrosis?	What are multiple alleles?	What is the Human Genome Project?	What is a restriction enzyme?	What is a transgenic organism?
25	What is Tay-Sachs disease?	What is Rh factor?	What is gel electrophoresis?	What is a sticky end?	What is a DNA fingerprint?
BONUS	What is phenylketonuria (PKU)?	What is genetic counseling?	What is gene therapy?	What is polymerase chain reaction (PCR)?	Who are Francis Collins and Craig Venter?

Evolution
Section 15

	Earth's History I	Earth's History II	Origin of Life	Evolution I	Evolution II
5	What is a fossil?	Who is Alexander Oparin?	What is spontaneous generation?	Who is Charles Darwin?	Who is Thomas Malthus?
10	What is a paleontologist?	Who are Stanley Miller and Harold Urey?	What is biogenesis?	What is H.M.S. *Beagle*?	What is artificial selection (or selective breeding)?
15	What is relative dating?	What are microspheres?	Who is Francesco Redi?	What are the Galapagos Islands?	What is fitness?
20	What is radioactive dating?	What is RNA?	Who is Lazzaro Spallanzani?	Who is Alfred Russell Wallace?	What is an adaptation?
25	What is a half-life?	What is ozone?	Who is Louis Pasteur?	Who is Jean Baptiste de Lamarck?	What is natural selection (or survival of the fittest)?
BONUS	What is Precambrian?	What is the endosymbiotic theory?	What is evolution?	Who is Charles Lyell?	What are populations?

Evolution
Section 16

	Evolution III	Evolution IV	Evolution V	Evolution VI	Primates
5	What are homologous structures?	What is genetic equilibrium?	What is disruptive selection?	What is mass extinction?	What is an opposable thumb?
10	What are analogous structures?	What is the Hardy-Weinberg principle?	What is speciation?	What is coevolution?	What is bipedalism?
15	What are vestigial structures?	What is genetic drift?	What is reproductive isolation?	What is adaptive radiation?	What are the prosimians?
20	What is descent with modification?	What is founder effect?	What is geographic isolation?	What is convergent evolution?	What are the anthropoids?
25	What is a gene pool?	What is directional selection?	What is behavioral isolation?	What is punctuated equilibrium?	What is a prehensile tail?
BONUS	What is allele frequency?	What is stabilizing selection?	What is temporal isolation?	What is gradualism?	What are the Great Apes?

Evolution
Section 17 (5)

	Human Evolution	Classification I	Classification II	Classification III	Classification IV
5	What are the hominids?	What is taxonomy?	What is a phylum?	What is a phylogenetic tree?	What is Animalia?
10	What are the australopithecines?	What is binomial nomenclature?	What is a class?	What are derived characteristics?	What is Plantae?
15	What is Lucy?	Who is Carolus Linnaeus?	What is an order?	What is a cladogram?	What is Fungi?
20	What is *Homo habilis*?	What is a dichotomous key?	What is a family?	What is cladistics?	What is Protista?
25	What is *Homo erectus*?	What is a kingdom?	What is a genus?	What is a domain?	What is Bacteria (or Eubacteria)?
BONUS	Who are the Cro-Magnons?	What is a division?	What is a species?	What is Eukarya?	What is Archaebacteria?

Microorganisms and Fungi
Section 18 (6)

	Bacteria Classification	Characteristics of Bacteria I	Characteristics of Bacteria II	Characteristics of Bacteria III	Viruses I
5	What is Archaebacteria?	What are cocci?	What are photoautotrophs?	What is binary fission?	What is a capsid?
10	What is Eubacteria?	What are bacilli?	What are saprobes?	What is conjugation?	What is a bacteriophage?
15	What are Gram-positive bacteria?	What are spirilla?	What are chemoautotrophs?	What is an endospore?	What is the lytic cycle?
20	What are Gram-negative bacteria?	What are streptococci?	What are obligate aerobes?	What is an antibiotic?	What is the lysogenic cycle?
25	What are methanogens?	What are staphylococci?	What are obligate anaerobes?	What is a pathogen?	What is a prophage?
BONUS	What are thermoacidophiles?	What are flagella?	What are facultative anaerobes?	What is nitrogen fixation?	What are glycoproteins?

Microorganisms and Fungi
Section 19

	Viruses II	Protist Phyla I	Protist Phyla II	Protist Characteristics I	Protist Characteristics II
5	What is icosahedron?	What is phylum Zoomastigina?	What is phylum Chlorophyta?	What are pseudopodia?	What is a holdfast?
10	What are retroviruses?	What is phylum Ciliophora?	What is phylum Rhodophyta?	What is a contractile vacuole?	What is agar?
15	What is human immunovirus (HIV)?	What is phylum Euglenophyta?	What is phylum Phaeophyta?	What is the gullet?	What is bioluminescence?
20	What are oncogenic viruses?	What is phylum Sporozoa?	What is phylum Bacillariophyta?	What is an eyespot?	What is plasmodium?
25	What are viroids?	What is phylum Sarcodines?	What is phylum Chrysophyta?	What is a micronucleus?	What is alternation of generations?
BONUS	What are prions?	What is the phylum Pyrrophyta (or Dinoflagellata)?	What is phylum Acrasiomycota?	What is a pellicle?	What are water molds?

Microorganisms and Fungi
Section 20

	Protist Characteristics III	Fungi I	Fungi II	Fungi III	Potpourri
5	What are food vacuoles?	What is Ascomycota?	What is chitin?	What is yeast?	What are vaccines?
10	What is a blade?	What are Zygomycetes?	What are hyphae?	What is a rhizoid?	What is plankton?
15	What is a trichocyst?	What is Basidiomycota?	What is a septum?	What is a stolon?	What is penicillin?
20	What is carageenan?	What is Deuteromycota?	What is a mycelium?	What are asci?	What is athlete's foot?
25	What is conjugation?	What is a lichen?	What is a sporangium?	What are conidiophores?	What is *Volvox*?
BONUS	What is a plasmodium?	What is a mycorrhiza?	What is a gametangium?	What are basidia?	What is botulism?

Plants
Section 21

	Plant Phyla I	Plant Phyla II	Mosses and Their Relatives	Ferns and Their Relatives	Plant Cells and Tissues I
5	What is Bryophyta?	What is Pterophyta?	What is a capsule?	What is a frond?	What is dermal tissue?
10	What is Hepaticophyta?	What is Coniferophyta?	What is a rhizoid?	What is a rhizome?	What are epidermal cells?
15	What is Anthocerophyta?	What is Ginkgophyta?	What is an antheridium?	What is a sporangium?	What is the cuticle?
20	What is Lycophyta?	What is Cycadophyta?	What is an archegonium?	What is a sorus?	What is meristematic tissue?
25	What is Psilotophyta?	What is Gnetophyta?	What is a protonema?	What is a prothallus?	What is an apical meristem?
BONUS	What is Sphenophyta (or Arthrophyta)?	What is Anthophyta?	What is a gemma?	What is a strobilus?	What is lateral meristem?

Plants
Section 22

	Plant Cells and Tissues II	Plant Cells and Tissues III	Roots	Stems I	Stems II
5	What is vascular tissue?	What is a companion cell?	What is a root cap?	What is a bud?	What is bark?
10	What is xylem?	What is a sieve plate?	What is a taproot?	What is a node?	What is heartwood?
15	What is phloem?	What is ground tissue?	What are fibrous roots?	What is an internode?	What is sapwood?
20	What is a tracheid?	What are parenchyma cells?	What are root hairs?	What is pith?	What are annual rings?
25	What is a vessel element?	What are collenchyma cells?	What is the cortex?	What is vascular cambium?	What is a tuber?
BONUS	What is a sieve-tube element?	What are sclerenchyma cells?	What is the endodermis?	What is cork cambium?	What is a bulb?

7 Plants
Section 23

	Leaves	Flowers	Reproduction in Seed Plants I	Reproduction in Seed Plants II	Seeds and Fruit I
5	What is a blade?	What are petals?	What is a pollen cone (or male cone)?	What is nectar?	What is a seed coat?
10	What is a petiole?	What are sepals?	What is a seed cone (or female cone)?	What is wind pollination?	What is a fruit?
15	What is a vein?	What is a stamen?	What is a microspore mother cell?	What is endosperm?	What is a cotyledon?
20	What is palisade mesophyll?	What is the pistil or carpel?	What is a megaspore mother cell?	What is double fertilization?	What is a monocotyledon?
25	What is spongy mesophyll?	What is the stigma?	What is a pollen tube?	What is vegetative reproduction?	What is a dicotyledon?
BONUS	What is a stoma?	What is the ovary?	What is a micropyle?	What is grafting?	What is germination?

7 Plants
Section 24

	Seeds and Fruit II	Plant Hormones	Plant Responses I	Plant Responses II	Potpourri
5	What is a radicle?	What is auxin?	What is a tropism?	What is abscission?	What are cultivars?
10	What is a hypocotyl?	What is ethylene?	What is phototropism?	What is photoperiodism?	What are xerophytic plants?
15	What is an epicotyl?	What is gibberellin?	What is gravitropism?	What is phytochrome?	What is an epiphyte?
20	What is dormancy?	What is cytokinin?	What is thigmotropism?	What is a short-day plant?	What is an annual?
25	What is a legume?	What is a herbicide?	What is chemotropism?	What is a long-day plant?	What is a biennial?
BONUS	What is a drupe?	What is abscisic acid?	What is thermotropism?	What is a day-neutral plant?	What is a perennial?

Invertebrates
Section 25

	Characteristics of the Animals I	Characteristics of the Animals II	Characteristics of the Animals III	Characteristics of the Animals IV	Characteristics of the Animals V
5	What is an invertebrate?	What is a blastula?	What is anterior?	What is an open circulatory system?	What is a filter feeder?
10	What is a vertebrate?	What is the endoderm?	What is posterior?	What is a closed circulatory system?	What is intracellular digestion?
15	What is an exoskeleton?	What is mesoderm?	What is dorsal?	What is acoelomate?	What is internal fertilization?
20	What is an endoskeleton?	What is ectoderm?	What is ventral?	What is pseudocoelomate?	What is external fertilization?
25	What is segmentation?	What is a protostome?	What is radial symmetry?	What is coelomate?	What is a hermaphrodite?
BONUS	What is a hydrostatic skeleton?	What is a deuterostome?	What is bilateral symmetry?	What is cephalization?	What is a larva?

Invertebrates
Section 26

	Invertebrate Classification I	Invertebrate Classification II	Sponges	Cnidarians and Ctenophores I	Cnidarians and Ctenophores II
5	What is Porifera?	What is Annelida?	What is sessile?	What is a polyp?	What is a nerve net?
10	What is Cnidaria?	What is Mollusca?	What is a choanocyte (or a collar cell)?	What is a medusa?	What are ocelli?
15	What is Ctenophora?	What is Echinodermata?	What is an osculum?	What is mesoglea?	What are colloblasts?
20	What is Platyhelminthes?	What is Arthropoda?	What are spicules?	What is a gastrovascular cavity?	What is bioluminescence?
25	What is Rotifera?	What is Chelicerata?	What are amoebocytes (or archaeocyte)?	What are cnidocytes?	What is Hydrozoa?
BONUS	What is Nematoda?	What is Crustacea?	What are gemmules?	What are nematocysts?	What are coral reefs?

Invertebrates
Section 27

	Flatworms I	Flatworms II	Rotifers and Roundworms	Annelids I	Annelids II
5	What is a pharynx?	What is Turbellaria?	What are cilia?	What are setae?	What is the crop?
10	What are flame cells?	What is Trematoda?	What is the mastax?	What are parapodia?	What are aortic arches?
15	What are ganglia?	What is Cestoda?	What are hookworms?	What are septa?	What is Oligochaeta?
20	What is an eyespot?	What is a scolex?	What is trichinosis?	What are nephridia?	What is Hirudinea?
25	What is fission?	What are proglottids?	What is elephantiasis?	What is a clitellum?	What is a proboscis?
BONUS	What is acoelomate?	What is a cyst?	What is *Caenorhabditis elegans*?	What are castings?	What is Polychaeta?

Invertebrates
Section 28

	Mollusks I	Mollusks II	Arthropod Characteristics I	Arthropod Characteristics II	Crustaceans I
5	What is a foot?	What is a siphon?	What is an appendage?	What is an open circulatory system?	What is nauplius?
10	What is the visceral mass?	What is Gastropoda?	What is chitin?	What are tracheal tubes?	What are cirri?
15	What is the mantle?	What is Bivalvia?	What is molting?	What is a spiracle?	What are copepods?
20	What is a shell?	What is an adductor muscle?	What is a compound eye?	What are book lungs?	What are isopods?
25	What is a radula?	What is Cephalopoda?	What are trilobites?	What are Malphigian tubules?	What are decapods?
BONUS	What is a trochophore?	What is a chambered nautilus?	What is the Burgess Shale?	What are horseshoe crabs?	What is the cephalothorax?

8. Invertebrates
Section 29

	Crustaceans II	Chelicerates and Uniramians	Insects I	Insects II	Echinoderms
5	What is the abdomen?	What are chelicerae?	What is three?	What is pollination?	What is a water vascular system?
10	What is the carapace?	What are pedipalps?	What is the labrum?	What is incomplete metamorphosis?	What is a ring canal?
15	What are antennae?	What are spinnerets?	What is the labium?	What is complete metamorphosis?	What is a tube foot?
20	What are mandibles?	What is Arachnida?	What is the gastric cecum?	What is a chrysalis?	What is a madreporite?
25	What are chelipeds?	What are millipedes?	What is the seminal receptacle?	What is a pheromone?	What are ossicles?
BONUS	What are swimmerets?	What are centipedes?	What is an ovipositor?	What are social insects?	What is Aristotle's lantern?

9. Vertebrates
Section 30

	Chordates and Nonvertebrate Chordates	Vertebrate Classification	Fishes I	Fishes II	Fishes III
5	What is a dorsal nerve cord?	What is Mammalia?	What is Agnatha?	What are scales?	What is a swim bladder?
10	What are pharyngeal pouches?	What is Aves?	What is a hagfish?	What is a gill?	What is oviparous?
15	What is a notochord?	What is Reptilia?	What is a lamprey?	What is the operculum?	What is ovoviviparous?
20	What is a postanal tail?	What is Amphibia?	What is the sinus venosus?	What is a caudal fin?	What is viviparous?
25	What is a lancet?	What is Osteichthyes?	What is single-loop circulation?	What is a pectoral girdle?	What is anadromous?
BONUS	What is a tunicate?	What is Chondrichthyes?	What is countercurrent flow?	What is a lateral line system?	What is catadromous?

Vertebrates
Section 31

	Amphibians I	Amphibians II	Reptiles I	Reptiles II	Birds I
5	What are tadpoles?	What is the cloaca?	What is a yolk sac?	What is keratin?	What is a beak or bill?
10	What are mucous glands?	What is a nictitating membrane?	What is the amnion?	What are dinosaurs?	What is a wing?
15	What is cutaneous respiration?	What are tympanic membranes?	What is the allantois?	What is Crocodilia?	What is a down feather?
20	What is three?	What is Anura?	What is Jacobson's organ?	What is Chelonia?	What is a contour feather?
25	What is amplexus?	What is Urodela?	What is carapace?	What is Squamata?	What are hollow bones?
BONUS	What is a urostyle?	What is Apoda?	What is plastron?	What is Rhynchocephalia?	What is Archaeopteryx?

Vertebrates
Section 32

	Birds II	Bird Orders	Mammals I	Mammals II	Mammal Orders
5	What is a gizzard?	What is Passeriformes?	What are mammary glands?	What is four?	What are Primates?
10	What is the sternum?	What is Falconiformes?	What is an endotherm?	What is echolocation?	What is Carnivora?
15	What are the air sacs?	What is Piciformes?	What is a diaphragm?	What is a monotreme?	What is Rodentia?
20	What is preening?	What is Galliformes?	What is the rumen?	What is a marsupial?	What is Cetacea?
25	What is migration?	What is Strigiformes?	What are incisors?	What is a placental mammal?	What is Chiroptera?
BONUS	What is a brood patch?	What is Anseriformes?	What are molars?	What are ungulates?	What is Artiodactyla?

Human Biology
Section 33

	Tissues and Organ Systems I	Tissues and Organ Systems II	The Skeleton I	The Skeleton II	Bone Structure
5	What is the skeletal system?	What is the digestive system?	What is the axial skeleton?	What is the skull?	What is spongy bone?
10	What is the muscular system?	What is the excretory system?	What is the appendicular skeleton?	What is the humerus?	What is compact bone?
15	What is epithelial tissue?	What is the nervous system?	What is the femur?	What is the sternum?	What is red marrow?
20	What is the integumentary system?	What is connective tissue?	What are the phalanges?	What is the tibia or fibula?	What is yellow marrow?
25	What is the respiratory system?	What is the reproductive system?	What is the knee?	What is the clavicle?	What are Haversian canals?
BONUS	What is the circulatory (or cardiovascular) system?	What is the endocrine system?	What is 206?	What is the wrist?	What is the periosteum?

Human Biology
Section 34

	Bone Formation	Joints	The Muscular System	Muscle Movement	The Integumentary System
5	What is cartilage?	What is a ball-and-socket joint?	What is cardiac muscle?	What is a tendon?	What is the epidermis?
10	What is ossification?	What is an immovable or fixed joint?	What is skeletal muscle?	What is the origin?	What is the dermis?
15	What are osteocytes?	What is a ligament?	What is smooth muscle?	What is the insertion?	What is a hair follicle?
20	What are osteoblasts?	What is synovial fluid?	What is myosin?	What is a flexor?	What is keratin?
25	What are osteoclasts?	What are bursa?	What is actin?	What is an extensor?	What is melanin?
BONUS	What is the epiphyseal plate (or growth plate)?	What is arthritis?	What is a sarcomere?	What is a neuromuscular junction?	What is a sebaceous gland?

Human Biology
Section 35

	The Digestive System I	The Digestive System II	The Digestive System III	The Excretory System I	The Excretory System II
5	What is the mouth?	What is the stomach?	What is the liver?	What are the kidneys?	What is a nephron?
10	What are the salivary glands?	What is chyme?	What is bile?	What is the urinary bladder?	What is the glomerulus?
15	What is the esophagus?	What is pepsin?	What is the gallbladder?	What is the ureter?	What is the Bowman's capsule?
20	What is a bolus?	What is the pyloric valve?	What is the pancreas?	What is the urethra?	What is the loop of Henle?
25	What is peristalsis?	What is the duodenum?	What is the large intestine (or colon)?	What is the renal cortex?	What is filtration?
BONUS	What is the cardiac sphincter?	What are villi?	What are bacteria?	What is urea?	What is reabsorption?

Human Biology
Section 36

	Food and Nutrition	The Human Heart I	The Human Heart II	Blood Vessels	Blood
5	What is a calorie?	What is four?	What is the aorta?	What are capillaries?	What is plasma?
10	What are carbohydrates?	What are the atria?	What is the pacemaker (or sinoatrial node)?	What are arteries?	What are red blood cells or erythrocytes?
15	What are fats (or lipids)?	What are the ventricles?	What is pulmonary circulation?	What are veins?	What are white blood cells (or leukocytes)?
20	What are proteins?	What is a valve?	What is systemic circulation?	What is blood pressure?	What is hemoglobin?
25	What are vitamins?	What is the septum?	What is the superior vena cava?	What is systolic pressure?	What are platelets?
BONUS	What are minerals?	What is the pericardium?	What are pulmonary veins?	What is diastolic pressure?	What is fibrinogen?

Human Biology
Section 37

	The Circulatory System and Blood Disorders	Blood Types	The Lymphatic System	The Respiratory System I	The Respiratory System II
5	What is a heart attack?	What are antigens?	What is lymph?	What is the pharynx?	What is oxygen?
10	What is atherosclerosis?	What is type AB blood?	What are lymph nodes?	What is the trachea?	What is carbon dioxide?
15	What is hypertension?	What is type O blood?	What are lymphocytes?	What is the epiglottis?	What is the diaphragm?
20	What is a stroke?	What are antibodies?	What is the spleen?	What is the larynx?	What is carbon monoxide?
25	What is hemophilia?	What is agglutination?	What is edema?	What are the bronchi?	What is bronchitis?
BONUS	What is leukemia?	What is Rh factor?	What is the superior vena cava?	What are the alveoli?	What is emphysema?

Human Biology
Section 38

	The Nervous System I	The Nervous System II	The Neuron	The Brain	The Senses I
5	What is the central nervous system?	What is a reflex?	What is the cell body?	What is the cerebrum?	What is a thermoreceptor?
10	What is the peripheral nervous system?	What is a synapse?	What is a dendrite?	What is the cerebellum?	What is a chemoreceptor?
15	What is the somatic nervous system?	What are neurotransmitters?	What is the axon?	What is the brain stem?	What is the pupil?
20	What is the autonomic nervous system?	What is the resting potential?	What is the myelin sheath?	What is the corpus callosum?	What is the tympanum (or eardrum)?
25	What is the sympathetic nervous system?	What is an action potential?	What is a sensory neuron?	What are the meninges?	What is a cone?
BONUS	What is the parasympathetic nervous system?	What is the refractory period?	What is a motor neuron?	What is the occipital lobe?	What is the semicircular canal?

Human Biology
Section 39

	The Senses II	Drugs and the Nervous System	The Endocrine System	Endocrine Glands	Hormones I
5	What is a mechanoreceptor?	What is addiction?	What is a hormone?	What is the pituitary gland?	What is testosterone or androgens?
10	What is a cornea?	What is a stimulant?	What is an exocrine gland?	What is the thyroid?	What is thyroxine?
15	What is the cochlea?	What is a depressant?	What is an endocrine gland?	What is the pancreas?	What is insulin?
20	What is the olfactory receptor (or olfactory bulb)?	What is fetal alcohol syndrome?	What is a steroid hormone?	What is the pineal gland?	What is epinephrine or norepinephrine (or adrenaline and noradrenaline)?
25	What is the optic nerve?	What is cocaine?	What is a nonsteroid hormone (or an amino acid-based hormone)?	What are the adrenal glands?	What is an antidiuretic hormone?
BONUS	What is the vitreous humor?	What are opiates?	What are prostaglandins?	What is the thymus?	What is oxytocin?

Human Biology
Section 40

	Hormones II	Endocrine System Disorders	The Male Reproductive System	The Female Reproductive System	The Menstrual Cycle
5	What is estrogen or progesterone?	What is diabetes mellitus?	What are sperm?	What is the ovary?	What is the follicular phase?
10	What is glucagon?	What is hypoglycemia?	What is the penis?	What is the vagina?	What is ovulation?
15	What is melatonin?	What is hyperthyroidism?	What are the testes?	What is the uterus?	What is the luteal phase?
20	What is parathyroid hormone?	What is hypothyroidism?	What is semen?	What is the Fallopian tube?	What is menstruation?
25	What is calcitonin?	What is goiter?	What is the epididymis?	What is the cervix?	What is menopause?
BONUS	What is prolactin?	What is cretinism?	What is the vas deferens?	What are fimbriae?	What is follicle-stimulating hormone (FSH) or luteinizing hormone (LH)?

Human Biology
Section 41

	Fertilization and Development I	Fertilization and Development II	The Immune System I	The Immune System II	Cancer
5	What is fertilization?	What is a fetus?	What is a pathogen?	What is an allergy?	What is a tumor?
10	What is a zygote?	What is the umbilical cord?	What is a fever?	What is a histamine?	What is a benign tumor?
15	What is a morula?	What is the amniotic sac?	What is an inflammatory response?	What is interferon?	What is a malignant tumor?
20	What is a blastocyst?	What is the placenta?	What is an immune response?	What is immunity?	What is metastasis?
25	What is implantation?	What is the chorion?	What is a cell-mediated immune response?	What is a vaccination?	What is a carcinogen?
BONUS	What is gastrulation?	What is labor?	What is a humoral immune response?	What is an autoimmune disease?	What is chemotherapy?

Share Your Bright Ideas

We want to hear from you!

Your name_____ Date_____

School name_____

School address_____

City_____ State_____ Zip_____ Phone number (_____)_____

Grade level(s) taught_____ Subject area(s) taught_____

Where did you purchase this publication?_____

In what month do you purchase a majority of your supplements?_____

What moneys were used to purchase this product?

____School supplemental budget   ____Federal/state funding   ____Personal

Please "grade" this Walch publication in the following areas:

Quality of service you received when purchasing .... A B C D
Ease of use... A B C D
Quality of content.. A B C D
Page layout... A B C D
Organization of material.. A B C D
Suitability for grade level... A B C D
Instructional value... A B C D

COMMENTS:_____
_____

What specific supplemental materials would help you meet your current—or future—instructional needs?
_____

Have you used other Walch publications? If so, which ones?_____

May we use your comments in upcoming communications?   ____Yes   ____No

Please **FAX** this completed form to **888-991-5755,** or mail it to

Customer Service, Walch Publishing, P. O. Box 658, Portland, ME 04104-0658

We will send you a **FREE GIFT** in appreciation of your feedback.   **THANK YOU!**